"Bloom has written an excellent overview of the main philosophical issues facing Christians working in the natural sciences. He shows that not only is there no conflict between science and belief in God, but there are now many scientific discoveries that support such belief. A clear, concise, and highly readable treatment. I highly recommend it, especially for college students."

Stephen C. Meyer, Director, Center for Science and Culture, Discovery Institute; author, *Signature in the Cell: DNA and the Evidence for Intelligent Design*

"With PhDs in Ancient Near Eastern Studies and Physics, John Bloom is one of the top thinkers today on the relationship between science and Christianity. Written with a clarity of style and level of approach that a freshman in college would have no trouble reading, Bloom traces the relationship between Christianity and science through history up to the present. This enables him to put his finger on the erroneous tension points between the two, debunk these alleged tension points, and chart a way forward. Though a small book, it is packed with important ideas and information. It is must reading for any college course in science and Christianity."

J. P. Moreland, Distinguished Professor of Philosophy, Biola University; author, *The Soul: How We Know It's Real and Why It Matters*

"With doctorates in physics and theology, John Bloom presents a biblically sound understanding of science as it relates to Christian faith, offering many intriguing historical insights along the way."

William A. Dembski, Senior Fellow, Center for Science and Culture, Discovery Institute; author, *Being as Communion*

THE NATURAL SCIENCES

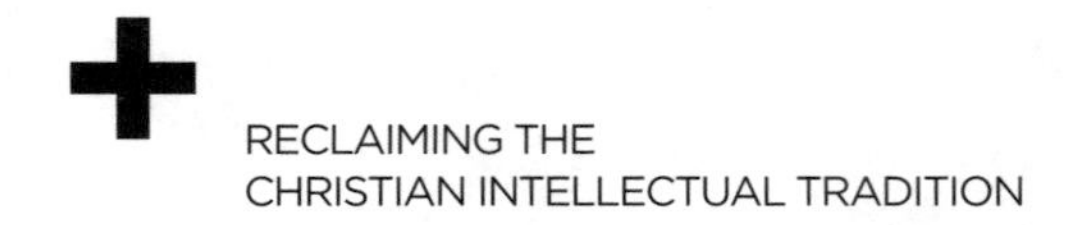

David S. Dockery, series editor

CONSULTING EDITORS

OTHER RCIT VOLUMES:

The Great Tradition of Christian Thinking, David S. Dockery and Timothy George

The Liberal Arts, Gene C. Fant Jr.

Political Thought, Hunter Baker

Literature, Louis Markos

Philosophy, David K. Naugle

Christian Worldview, Philip G. Ryken

Art and Music, Paul Munson and Joshua Farris Drake

Ethics and Moral Reasoning, C. Ben Mitchell

THE NATURAL SCIENCES
A STUDENT'S GUIDE

John A. Bloom

CROSSWAY
WHEATON, ILLINOIS

The Natural Sciences: A Student's Guide

Published by Crossway
1300 Crescent Street
Wheaton, Illinois 60187

Cover design: Jon McGrath, Simplicated Studio

First printing 2015

Printed in the United States of America

All emphases in Scripture quotations have been added by the author.

Trade paperback ISBN: 978-1-4335-3935-0
ePub ISBN: 978-1-4335-3938-1
PDF ISBN: 978-1-4335-3936-7
Mobipocket ISBN: 978-1-4335-3937-4

Library of Congress Cataloging-in-Publication Data

Bloom, John A., 1952-
The natural sciences : a student's guide / John A. Bloom.
pages cm. — (Reclaiming the Christian intellectual tradition)
Includes bibliographical references and index.
ISBN 978-1-4335-3935-0 (tp)
1. Religion and science. 2. Christian philosophy. 3. Naturalism—Religious aspects—Christianity. 4. Science—Philosophy. I. Title.
BL240.3.B625 2015
261.5'5—dc23 2014028257

Crossway is a publishing ministry of Good News Publishers.

VP 25 24 23 22 21 20 19 18 17 16 15
15 14 13 12 11 10 9 8 7 6 5 4 3 2 1

To Claudia,
my wife,
helpmate,
and best friend forever.

CONTENTS

SERIES PREFACE

RECLAIMING THE CHRISTIAN INTELLECTUAL TRADITION

The Reclaiming the Christian Intellectual Tradition series is designed to provide an overview of the distinctive way the church has read the Bible, formulated doctrine, provided education, and engaged the culture. The contributors to this series all agree that personal faith and genuine Christian piety are essential for the life of Christ followers and for the church. These contributors also believe that helping others recognize the importance of serious thinking about God, Scripture, and the world needs a renewed emphasis at this time in order that the truth claims of the Christian faith can be passed along from one generation to the next. The study guides in this series will enable us to see afresh how the Christian faith shapes how we live, how we think, how we write books, how we govern society, and how we relate to one another in our churches and social structures. The richness of the Christian intellectual tradition provides guidance for the complex challenges that believers face in this world.

This series is particularly designed for Christian students and others associated with college and university campuses, including faculty, staff, trustees, and other various constituents. The contributors to the series will explore how the Bible has been interpreted in the history of the church, as well as how theology has been formulated. They will ask: How does the Christian faith influence our understanding of culture, literature, philosophy, government, beauty, art, or work? How does the Christian intellectual tradition help us understand truth? How does the Christian intellectual tradition shape our approach to education? We believe that this series is not only timely but that it meets an important need, because the secular culture in which we now find ourselves is, at best, indifferent to the Christian faith, and the Christian world—at least in its more popular forms—tends to be confused about the beliefs, heritage, and tradition associated with the Christian faith.

At the heart of this work is the challenge to prepare a generation of Christians to think Christianly, to engage the academy and the culture, and to serve church and society. We believe that both the breadth and the depth of the Christian intellectual tradition need to be reclaimed, revitalized, renewed, and revived for us to carry forward this work. These study guides will seek to provide a framework to help introduce students to the great tradition of Christian thinking, seeking to highlight its importance for understanding the world, its significance for serving both church and society, and its application for Christian thinking and learning. The series is a starting point for exploring important ideas and issues such as truth, meaning, beauty, and justice.

We trust that the series will help introduce readers to the apostles, church fathers, Reformers, philosophers, theologians, historians, and a wide variety of other significant thinkers. In addition to well-known leaders such as Clement, Origen, Augustine, Thomas Aquinas, Martin Luther, and Jonathan Edwards, readers will be pointed to William Wilberforce, G. K. Chesterton, T. S. Eliot, Dorothy Sayers, C. S. Lewis, Johann Sebastian Bach, Isaac Newton, Johannes Kepler, George Washington Carver, Elizabeth Fox-Genovese, Michael Polanyi, Henry Luke Orombi, and many others. In doing so, we hope to introduce those who throughout history have demonstrated that it is indeed possible to be serious about the life of the mind while simultaneously being deeply committed Christians. These efforts to strengthen serious Christian thinking and scholarship will not be limited to the study of theology, scriptural interpretation, or philosophy, even though these areas provide the framework for understanding the Christian faith for all other areas of exploration. In order for us to reclaim and advance the Christian intellectual tradition, we must have some understanding of the tradition itself. The volumes in this series seek to explore this tradition and its application for our twenty-first-century world. Each volume contains a glossary, study questions, and a list of resources for further study, which we trust will provide helpful guidance for our readers.

I am deeply grateful to the series editorial committee: Timothy George, John Woodbridge, Michael Wilkins, Niel Nielson, Philip Ryken, and Hunter Baker. Each of these colleagues joins me in thanking our various contributors for their fine work. We all express our appreciation to Justin Taylor, Jill Carter, Allan Fisher, Lane Dennis, and the Crossway team for their enthusiastic support for the project. We offer the project with the hope that students will be helped, faculty and Christian leaders will be encouraged, institutions will be strengthened, churches will be built up, and, ultimately, that God will be glorified.

Soli Deo Gloria
David S. Dockery
Series Editor

AUTHOR'S PREFACE

I ask God to make my spirit strong so that I direct my glance at the pure truth, from whichever side it should be presented, and do not let myself be misled, as so often happens today, by the admiration or contempt of persons or sides.

Johannes Kepler, 1606[1]

If you love science, it's likely that you avoid philosophy and history and that you will try to limit your exposure in these areas to a couple of the easiest general education courses. That's certainly what I and most of my science-minded friends did when we were in college. To some extent this is justifiable, since kings, battles, Kant, and Sartre are very different from cells, lasers, rockets, and computers. But scientists think. And thinking involves certain assumptions. And those assumptions have changed over time. Thus philosophy (reflecting on our assumptions) and history (where our current assumptions come from) are important for scientists to consider so that we can think better.

Many scientists today are technically competent, but because of their high specialization, they are not aware of and do not reflect upon the foundations of their discipline. For you as a Christian student, aspiring to and training for a career in a scientific field, it is important to be counter-cultural in this regard so that you can face the prevailing winds in our society, which presume that "smart people do not believe in God anymore," or that "Christianity and science are always at war with each other," and similar myths.

My hope is that this brief introduction to the rich Christian

[1] Max Casper, *Kepler*, trans. and ed. C. Doris Hellman (New York: Dover, 1993), 373. Quotation from *Syla chronologica* (part 4 of *De stella nova*).

intellectual tradition in the sciences will spur you to make this area a hobby (at least!) of lifelong interest.[2] The richness of knowing why we think what we do and of reflecting on God's handiwork in the world around us gives a depth and life to science that makes it all the more fascinating and rewarding.

[2] During my long years as a graduate student, I supported myself doing computer consulting and got involved in a project which sought input from Dr. Rocco "Rocky" Martino, a successful multimillionaire computer genius who had helped write the Waterloo Fortran compiler and developed disk-shadowing and other algorithms to secure banking-industry transactions. I felt a bit insignificant as a poor graduate student with a crazy idea meeting with the president and CEO of XRT, a major innovative software company at the time. Wondering about my background, Dr. Martino asked why I was pursuing a degree in Ancient Near Eastern Studies when I already had a PhD in Physics. I mentioned that my interests were on the interface of Christianity and science, on topics like creation and evolution. There was a long pause, and then he said, "Wow, that's a great area! If I could do it all over again, I would go into that too!" That wasn't the answer I was expecting, and I took great encouragement from Dr. Martino's words.

ACKNOWLEDGMENTS

I wish to thank David Dockery for inviting me to contribute to this important series, and I'm thankful also for helpful input from Mike Keas and other colleagues in the Science and Religion program here at Biola University. I thank Biola for enabling and encouraging me to teach a course for over ten years specifically on the integration of Christianity and the natural sciences, whose material has filtered into this book in various ways.

Looking back, I wish to thank those who enabled me to pursue a multi-graduate-degree background: the late Herman J. Eckelmann, pastor of the former Faith Bible Church in Ithaca, NY, who encouraged me and many others to get trained in theology and apologetics; Dr. Watt Webb, my advisor at Cornell, who did not think it entirely crazy that I wanted to go to seminary, and invited me to postdoc with him over the summers to help pay my bills; the members of the Bethany Bible Fellowship Church in Hatfield, PA, who provided a wonderful "work for rent" relationship which lasted a bit longer than we all initially expected (thirteen years!); my wife, family, and friends who have borne with the graduate student and busy professor lifestyle (is there much difference?) forever; and Laura Talcott and the other editors at Crossway for their help in polishing this work.

To God be the glory!

INTRODUCTION

I am fortunate to live in southern California, a few miles from "The Happiest Place on Earth," the Jet Propulsion Laboratory (JPL) in Pasadena. JPL is an incredible place to visit and marvel at the ingenuity and design of the fleet of spectacular space probes and Mars landers and rovers, which this lab has deployed throughout the solar system for decades. Disneyland, perhaps the better known "Happiest Place on Earth," is incredible too; there I marvel at the ingenuity and design of the rides and entertainment, which the Disney "Imagineers" made for millions to enjoy.

What puzzles me as a scientist is that I can openly marvel at the ingenuity and design of the things that brilliant engineers create, but after I leave JPL or Disneyland and look at the even more incredible ingenuity and design in the world around me, I am told that I must explain it all away as the result of unguided natural processes. Why is there a disconnect here? Why can't I see a designer of the world around me, who crafted things far more ingenious and complex than any human engineer can imagine?

Of course, as a Christian who is a scientist, I can. But this is the great Christian intellectual tradition that the modern sciences have abandoned. A Christian can see the heavens and the earth as the handiwork of a powerful and ingenious God, where we have the privilege, challenge, and joy "to think God's thoughts after Him"[1] as we study his creation. The secular scientist is allowed only to see a universe that somehow got here by accident without an ultimate explanation, that works without purpose or direction, and within which we are a fluke. This approach is well summarized

[1] Attributed to Johannes Kepler. *New World Encyclopedia*, s.v. "Johannes Kepler," last revised June 10, 2014, http://www.newworldencyclopedia.org/entry/Johannes_Kepler.

by theoretical physicist Steven Weinberg's famous comment: "The more the universe seems comprehensible, the more it also seems pointless."[2] Such a fatalistic and depressing view of nature and of ourselves is a far cry from one which sees the universe as a glorious masterpiece and an ingenious puzzle to explore and which sees ourselves as bearing the Creator's image.

The purpose of this short book is to give students an overview of the people and events which gave rise both to science itself and to this modern naturalistic perspective. Once we appreciate where we have come from, we can survey the current scene, look forward, and suggest a means to rekindle the original Christian vision of science.

We will begin by laying out what exactly is the Christian intellectual tradition in the sciences, drawing from biblical passages which tell us about God's role in and relationship to his creation, and showing how Scripture gave Christians the correct way to view the world so that scientific studies could flourish. Next we will look at how Aristotle and other Greek philosophers influenced the development of science, and we'll consider Johannes Kepler, the mathematician and astronomer who well exemplifies early science's excitement and expectation of seeing God's genius in nature.

Then we will see how God's hand in nature was brushed aside as the Enlightenment movement placed absolute trust in human reason and experience while disdaining ancient authorities. Encouraging this shift was the discovery of mathematical and mechanical models that could predict behavior very well in physics and chemistry. Since it seemed as if God governed the world through laws and wasn't as active in it as originally imagined, we will see how science shifted from a theistic to a deistic perspective—then to a completely naturalistic one—with the help of secularizing cultural pressures.

---

[2] Steven Weinberg, *The First Three Minutes: A Modern View of the Origin of the Universe*, 2nd ed. (New York: Basic, 1993), 154.

We will follow this naturalistic thread into the world of biology, where Charles Darwin and others used the problem of evil, clever rhetoric, and a naturalistic definition of science to neatly exclude God from this discipline.

After surveying this history, we will work on defining what exactly "science" is, what its goals are, and how naturalism impacts modern science. We will note that science today has two projects: (1) finding out how things work, a task that focuses on mechanical processes and is largely metaphysically neutral; and (2) asking about how things originated, a goal which is religiously loaded. Since contemporary science has chosen to restrict itself to giving naturalistic explanations for phenomena *no matter what*, we will find that the major tensions today in Christianity and science actually occur between Christianity and naturalism, two different religious positions, not between Christianity and data. This distinction is critical since most people respect science today for its objective and neutral pursuit of truth and are not aware of how its commitment to naturalism can seriously skew its results on questions of origins. Once we are aware of this, there is considerable common ground for those of differing worldviews to work together in the sciences by focusing on data and not on personal metaphysical interpretations. We will discuss the options of limiting science to making modest claims about physical cause and effect or of allowing science the metaphysical freedom to think outside of its current naturalistic box.

We will next survey the evidence which strongly suggests that the naturalistic box is inadequate: human reason and knowledge are fundamentally limited; our universe is so fine-tuned that it appears far more likely that we are simply hallucinating the world around us than that it actually exists; and "simple" cells turn out to be unimaginably complex, hardwired biological computational devices. Across multiple scientific disciplines it appears that the Christian intellectual tradition offers a better vantage point for viewing the world around us unless we simply dismiss the universe as "natural magic."

Finally, we will conclude with some pointers about how Christians can best lead the way as scientists in the future. Since a range of Christian positions (reflecting Aristotelian, Neoplatonic, mechanical, and other philosophical influences) stimulated fruitful dialogue in the past, a range of positions should continue to be beneficial as we move forward, spurring an ongoing discussion of key integrative and scientific questions. The Christian tradition's motivation for studying the world to glorify God and to help others with that knowledge can be tremendous salt and light to our secular culture.

 1

WHAT IS THE CHRISTIAN INTELLECTUAL TRADITION IN THE SCIENCES?

I believe in God the Father Almighty, Maker of heaven and earth.

The Apostle's Creed

The best place for us to start is by noting that there *is* a Christian intellectual tradition in the sciences. Popular myths about the relationship between science and religion would have you believe that Christianity and science always were, and still are, at war with each other—but historians tell us quite the opposite: Christianity furnished the fertile soil in which science developed and flourished. In fact, at one time the church was the major sponsor of scientific work. So how did Christianity nourish and support science? By providing the correct worldview.

LAYING THE RIGHT FOUNDATION: PERSPECTIVE

In order to study the world fruitfully, **you have to look at it the right way**. If you think the world is full of gods who are constantly squabbling with each other and need to be appeased or avoided, you have no expectation that the world will behave in a regular way. Such was the universal polytheistic/pantheistic view that we find throughout the ancient world. But the Judeo-Christian perspective changed that: if there is only one God, and he is sovereign over his creation, then the universe is not run by a fickle committee. Nor are the physical things of the world gods, or manifestations of the gods. Stuff is just stuff, and it need not be feared. Matter doesn't have a personality to be angry, sad, malicious, or cooperative depending on its mood. The Bible *depersonalizes* nature by describing the sun and moon as objects in Genesis 1, and it calms Israel's fears about possible gods in the sky:

> Thus says the LORD:
>
> "Learn not the way of the nations,
> nor be dismayed at the signs of the heavens
> because the nations are dismayed at them,
> for the customs of the peoples are vanity." (Jer. 10:2–3)

Moreover, God tells us in many passages that he set up the heavens and earth to work according to fixed patterns. Here are some examples:

> While the earth remains, seedtime and harvest, cold and heat, summer and winter, day and night, shall not cease. (Gen. 8:22)

> Thus says the LORD: If you can break my covenant with the day and my covenant with the night, so that day and night will not come at their appointed time, then also my covenant with David my servant may be broken, so that he shall not have a son to reign on his throne, and my covenant with the Levitical priests my ministers. (Jer. 33:20–21)

> Thus says the LORD: If I have not established my covenant with day and night and the fixed order of heaven and earth, then I will reject the offspring of Jacob and David my servant and will not choose one of his offspring to rule over the offspring of Abraham, Isaac, and Jacob. For I will restore their fortunes and will have mercy on them. (Jer. 33:25–26)

Thus the first pillar of the Christian intellectual tradition's foundation in science rests on God's creation being *impersonal* and following *regular patterns*. What a relief that there is only one Being whom we need to concern ourselves with (Deut. 6:13)—not a frenzied zoo of powers—and that he and his creation are not capricious!

As part of seeing the world the right way, one needs to have **the correct idea about time**. Ancient cultures commonly had static or cyclical notions of time: either everything stays the way it's always been or else things will eventually repeat themselves like the seasons do. But the Bible is rare, if not unique, in presenting a linear notion of time with a beginning, an unfolding story, and an end to that story. This proper sense of time instills the hope of progress and a value to learning, rather than a fatalistic resignation to whatever happens.[1]

LAYING THE RIGHT FOUNDATION: MOTIVATION

In order to study the world, **you need the motivation to do so**. If you think that physical stuff is evil or illusionary, then you focus your life on the spiritual world and mystical experience, as the Christian heresy of gnosticism once did and as many other world religions do today. But the Bible teaches that *the study of nature is a worthy pursuit to gain wisdom and glorify God*. God's creation is certainly corrupted by sin, but we are encouraged to learn from it as noted in many passages:

> Go to the ant, O sluggard;
> consider her ways, and be wise. (Prov. 6:6)

> It is the glory of God to conceal things,
> but the glory of kings is to search things out. (Prov. 25:2)

> But ask the beasts, and they will teach you;
> the birds of the heavens, and they will tell you;
> or the bushes of the earth, and they will teach you;

[1] John Warwick Montgomery, *The Shape of the Past*, rev. ed. (Minneapolis, MN: Bethany Fellowship, 1975).

and the fish of the sea will declare to you.
Who among all these does not know
that the hand of the LORD has done this? (Job 12:7–9)

The heavens declare the glory of God,
and the sky above proclaims his handiwork. (Ps. 19:1)

The founders of modern science saw themselves as glorifying God as they studied his handiwork. Galileo remarked in his letter to the Grand Duchess Christina, "The glory and greatness of Almighty God are marvelously discerned in all His works and divinely read in the open book of Heaven."[2] Thus a second pillar of the foundation rests on the *value of studying nature*: it is a good thing to do.

The Christian tradition brings an additional motivation to this foundational pillar beyond the abstract pursuit of wisdom: "The relief of man's estate."[3] Francis Bacon was among the first to encourage scientific studies, not only to glorify God, but in order to overcome some of the effects of the fall through better technology and medicine. This promise of a better future continues as a strong drive in the sciences and in our culture today, even in its secular context, but the application of scientific knowledge for practical and beneficial ends has its roots in *the Christian call to relieve suffering and to help others*.[4]

LAYING THE RIGHT FOUNDATION: APPROACH

In order to study the world, **you need to be patient.** Most of nature's regularities are subtle and confounded by multiple effects that are all happening at the same time. For example, it is difficult to see the conservation of momentum using objects much bigger

[2] Galileo Galilei, "*Letter to the Grand Duchess Christina of Tuscany, 1615*," *Fordham University Modern History Sourcebook*, August 1997, http://www.fordham.edu/halsall/mod/galileo-tuscany.asp.
[3] Francis Bacon, *The Advancement of Learning*, 1.5.11.
[4] Christopher B. Kaiser, *Creational Theology and the History of Physical Science: The Creationist Tradition from Basil to Bohr* (Leiden: Brill, 1997) has an excellent summary of early Christian medicine.

than individual atoms, because frictional forces obscure it. Years of work and study are necessary to develop the equipment and the mathematical tools that allow us to model what exactly is happening in the physical world. The Christian worldview teaches us, in an additional pillar of the foundation, that *hard work is good and satisfying*,[5] that our senses are generally reliable, and that our efforts are worthwhile because there is truth to be found.

In order to study the world, **you must expect the unexpected.** Philosophers call this the *contingency of nature*: God could have created the world any way he wanted to, so we can't figure out how it works by sitting in an armchair and applying human logic alone. God does whatever he pleases "in heaven and on earth, in the seas and all deeps" (Ps. 135:6, see also Ps. 115:3). Thus if we want to learn what God actually did and is doing, another pillar of the foundation is that we need *to go out and look at it*. Interestingly enough, many of the greatest discoveries in science were unexpected and came from doubting conventional wisdom. The physicist Richard Feynman once famously quipped, "Science is the belief in the ignorance of experts."[6]

In order to study the world, **you need to trust others.** While sometimes one can make progress by doubting the experts, no one has the time to repeat centuries of experimental and theoretical work, thus we must stand cautiously on the shoulders of our teachers and other scientists. For science to flourish, a community of trusted individuals must work together to share their insights and results. This requires, as the last pillars to the Christian intellectual tradition in the sciences, both *collegiality* and *high ethical standards*. Unfortunately, fudging or "selecting" data in order to

[5] Adam was created to cultivate and tend the garden of Eden *before the fall*; thus God intended work to be a good thing. After the fall, our work became harder and more frustrating, but nevertheless, work can and should be a fulfilling and satisfying aspect of our lives. A "do the minimum to get by, so that you have the most time to play" attitude is not a Christian approach to life. John Calvin in *The Golden Booklet of the True Christian Life* and the apostle Paul in many of his letters encouraged believers to pursue their work (even lowly manual labor) as God's calling in their lives and to glorify God through the work of their hands.

[6] Richard P. Feynman, "What is Science?" *The Physics Teacher* 7, no. 6 (September 1969): 313.

obtain grants or publish papers is reportedly becoming a serious ethical problem in the sciences today. For example, pharmaceutical companies have lost millions of dollars in attempting to develop promising new drugs that were initially reported to work, but the "successful" results could not be replicated later by others. As the ethical standards of scientists (like the rest of our culture) weaken, scientific progress becomes much harder, because it is difficult to trust anyone else's work.

In addition to these main pillars, there are subtle ones. Because the Christian intellectual tradition offers the most realistic picture of human nature, it provides the best cultural setting for science and technology to flourish for everyone's good. For example, as selfish as it sounds, the promise of *personal gain* motivates almost everyone to work harder. The ability for inventors to get rich is something that our society has learned to protect through patents and copyrights, because otherwise this powerful drive to improve and innovate is lost. While we normally think of selfishness as a bad thing, if it is properly harnessed and rewarded, it encourages people's creative drives in music, art, technology, and the sciences.

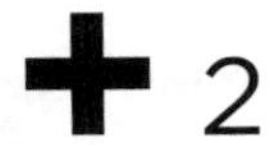

SEEING GOD IN THE DETAILS

In the beginning, God created the heavens and the earth.

Genesis 1:1

GOD AS CREATOR

From its opening verse, the Bible defines God as the Creator of the universe. This answer to the "Who is God?" question is repeated frequently in both the Old and New Testaments. Sample passages include:

> Thus says the LORD, your Redeemer,
> who formed you from the womb:
> "I am the LORD, who made all things,
> who alone stretched out the heavens,
> who spread out the earth by myself." (Isa. 44:24)

> All things were made through him, and without him was not any thing made that was made. (John 1:3)

> By faith we understand that the universe was created by the word of God, so that what is seen was not made out of things that are visible. (Heb. 11:3)

But note that these passages also allow us to infer something about our physical universe: it hasn't always been here. Early Jewish and Christian scholars argued repeatedly with pagan philosophers over this point. The common ancient view was that matter had always existed, but it developed into new forms as new generations of gods came into being. But since the Bible states that everything was created by God, theologians recognized early on that there could be nothing eternal besides God himself.

GOD'S CHARACTER

But we can make a stronger inference. By studying nature, we learn about the character of the artist behind it all:

> The heavens declare the glory of God,
> and the sky above proclaims his handiwork.
> Day to day pours out speech,
> and night to night reveals knowledge.
> There is no speech, nor are there words,
> whose voice is not heard.
> Their voice goes out through all the earth,
> and their words to the end of the world. (Ps. 19:1–4a)

> For his invisible attributes, namely, his eternal power and divine nature, have been clearly perceived, ever since the creation of the world, in the things that have been made. (Rom. 1:20a)

The apostle Paul argues here that everyone sees at least two attributes of God in his handiwork: (1) his power (given the size of the universe and the amount of energy it contains—this is awesome), and (2) his transcendence (God has to be bigger than the things he has made).

GOD'S IN CONTROL

But there is more. The Scriptures portray God as sovereign over his kingdom and as the designer, sustainer, and caregiver of his creation. Once again, some sample passages will give you the flavor:

> By the word of the LORD the heavens were made,
> and by the breath of his mouth all their host.
> He gathers the waters of the sea as a heap;
> he puts the deeps in storehouses.

Let all the earth fear the Lord;
let all the inhabitants of the world stand in awe of him!
For he spoke, and it came to be;
he commanded, and it stood firm. (Ps. 33:6–9)

Whatever the Lord pleases, he does,
in heaven and on earth,
in the seas and all deeps. (Ps. 135:6)

O Lord, how manifold are your works!
In wisdom have you made them all;
the earth is full of your creatures.
Here is the sea, great and wide,
which teems with creatures innumerable,
living things both small and great. (Ps. 104:24–25)

The Lord by wisdom founded the earth;
by understanding he established the heavens. (Prov. 3:19)

Is it by your understanding that the hawk soars
and spreads his wings toward the south?
Is it at your command that the eagle mounts up
and makes his nest on high? (Job 39:26–27)

GOD'S A GENIUS

The biblical picture which emerges sees heaven and earth not merely as God's kingdom, but as God's works of art and genius, exhibiting awe-inspiring wisdom and craftsmanship. Animals, plants, and the agrarian environment are often the focus of the biblical writers, as this was the world they were most familiar with, but today with microscopes and telescopes, we can feel an even deeper awe than King David:

When I look at your heavens, the work of your fingers,
the moon and the stars, which you have set in place,
what is man that you are mindful of him,
and the son of man that you care for him? (Ps. 8:3–4)

THE HEART AND SOUL OF THE TRADITION

Here is the heart and soul of the Christian intellectual tradition in the sciences: seeing the universe as the craftsmanship of God, the earth as his handiwork, living things as his special joys and marvels, and us as his stewards made "a little lower than the angels." With this perspective, the study of creation is the glorious pursuit of knowledge and wisdom that comes from thinking God's thoughts after him. We study the world around us as we might a great work of art like the *Mona Lisa*: looking for patterns, regularities, and style, to be sure, while also being sensitive to the artist's creativity, clever technique, and brilliant execution, which make the piece more than *simply* spatters on canvas or a chemical photograph.[1] God's creation is both a jewel and a puzzle to explore, so like Nicolaus Copernicus we can seek "the mechanism of the universe, wrought for us by a supremely good and orderly Creator."[2]

And if this isn't enough, there is a practical benefit to gaining this wisdom: we can use much of it to relieve suffering, help others, and improve our quality of life. As Genesis 1:28 and many passages in Proverbs promise, the fruit of our culture's pursuit of scientific wisdom allows the average first-world person today to enjoy the blessing of living healthier and far more comfortably than even the richest kings could live in the past. The Christian intellectual tradition sees science and technology as one of God's blessings, tempered by the awareness that sinful motives may distort it—just as sin can ruin any of God's other good gifts.

---

[1] See Benjamin Wiker and Jonathan Witt, *A Meaningful World: How the Arts and Sciences Reveal the Genius of Nature* (Westmont, IL: InterVarsity, 2006) for an excellent treatment revealing God's genius in nature.

[2] Nicolaus Copernicus, *On the Revolutions of the Heavenly Spheres*, preface.

EARLY SCIENCE

THE HANDMAIDEN TO THEOLOGY

God is the beginning and the end of scientific research and striving.

Max Casper, summarizing Johannes Kepler's thinking[1]

A TIME BEFORE SCIENCE

It may be hard to imagine, but there once was a time when science as we know it today did not exist. During the Roman empire, when Christianity itself was getting started, medicine was a simple and sometimes crude practice, mathematics was used chiefly for business and surveying, and astronomy was closely associated with astrology. Although the civil engineering of roads, buildings, and aqueducts had developed through trial-and-error experience, there were virtually no recognizable forms of modern chemistry, physics, or biology. The discussion of these topics, when it occurred at all, happened under the broad umbrella of philosophy and consisted mostly of simple observations and speculation about nature. It would be a mistake, however, to conclude that people back then were stupid and ignorant: classical Greek and Roman scholars put their intellectual efforts into philosophy and ethics because the systematic study of nature was not seen as practical, useful, or valuable. In general, pagan philosophies saw the physical world as capricious and encouraged their followers to avoid it as much as possible.

While nature itself was not studied in a detailed or systematic way, Greco-Roman philosophers nevertheless argued over various

[1] Max Casper, *Kepler*, trans. and ed. C. Doris Hellman (New York: Dover, 1993), 374.

scientific ideas and their related metaphysical issues: the structure of the world and heavens, how the world came to be, the significance of humans in it, and the gods' roles in the world. Thus the early Christians faced the challenge of living and presenting a biblical worldview that differed in important ways from the pagan ones. Like the Old Testament writers, apologists from the apostle Paul onward argued that the cosmos was intentionally created by one God, in contrast to the pagan ideas that matter had always been here, shaped by some demiurge (a subordinate deity) unconsciously or by procreating gods into the forms we now see. Augustine's arguments that the physical world is not evil, but that the human will is evil, helped refute the common gnostic tendency to see the world through a dualistic lens that equated the physical with evil and the spiritual with good.[2] Perhaps most stunning of all was the biblical view of humankind: people are not slaves charged with caring for and feeding the gods but are the benefactors of a totally self-sufficient God who invites them to become his friends—friends who are given the privilege of being his ambassadors and stewards over his creation, and who are fundamentally equal to each other.[3]

This Christian view about the origin and nature of the world gave people a deep motive to study it. By pondering the creation around us, we can marvel over God's handiwork and craftsmanship, and thus praise him for the things he has made, as is so often done in the Psalms. Thus early science in the West came to be viewed as the "handmaiden to theology." While the greatest human endeavor is the study of God himself through theology, the study of God's works also clearly points people to God and is thus a worthy pursuit.[4] When the modern university system of formal higher

---

[2] Augustine, *Confessions*, 4.15.24.

[3] Paul's address to the Athenians on Mars Hill (Acts 17:24–29) lays out concisely the contrasts between the Christian and Greco-Roman worldviews and religions.

[4] Johannes Kepler reflects this sentiment in a letter to a former professor: "I had the intention of becoming a theologian. For a long time I was restless: but now I see how God is, by my endeavors, also glorified in astronomy." Letter to Michael Maestlin, 3 October 1595, in Carola Baumgardt, *Johannes Kepler: Life and Letters* (New York: Philosophical Library, 1951), 31.

education was founded by the church in the 1100s, the systematic study of nature was given a stable platform within the academy. Prior to this, if practiced at all, it was mostly a hobby of the idle rich or of a few scholars whom the rich supported financially.

ARISTOTLE REDISCOVERED

With the rediscovery of Aristotle in the West, around the same time that universities were starting, the proper analytical tools to observe nature and quantify new knowledge became available—this time without the pagan bias regarding the material world. One of the important analytical tools which Aristotle presented was the four ways we can answer the "why" questions. Philosophers today categorize these as the "four causes":

- The *material* cause: that from which an item is made (e.g., the bronze of a statue)
- The *formal* cause: the form or pattern that makes an item what it is (e.g., the shape of a statue)
- The *efficient cause*: the primary source of the change or rest (e.g., the artisan, the art of bronze-casting the statue, the man who gives advice, or the father of the child)
- The *final* cause: the goal or purpose for which a thing is done (e.g., health is the goal of walking, losing weight, and surgical tools)[5]

While all four causes are used when studying nature, Aristotle argued that the final cause is the most important one. Thus Aristotelian philosophers focused primarily on the purpose or *teleology* behind physical events. While this approach worked well with biology questions ("Why does an acorn grow into an oak tree?"), the assumption that all objects had a personality, "soul," or some internal driving force (like the DNA in the acorn) did not work very well in physics (for example, "The rock falls because it wants to get

[5] Adapted from Andrea Falcon, "Aristotle on Causality," in *The Stanford Encyclopedia of Philosophy*, ed. Edward N. Zalta, last revision October 15, 2012, http://plato.stanford.edu/archives/win2012/entries/aristotle-causality/.

closer to the ground"). Ascribing motives and desires to inanimate objects was problematic, since in a Christian view such things were not gods or the manifestations of gods, nor endowed with souls.

COMPETING VIEWS

Besides Aristotle's approach, there were competing views from the classical Greek world, which also had a strong influence on the beginnings of science. *Neoplatonism* looked somewhat mystically for mathematical patterns in nature, and a Christianized version of this search for elegance and "active principles" inspired Nicolaus Copernicus and Johannes Kepler to propose and defend the heliocentric model of the solar system. In contrast to the Neoplatonists, the classical Epicurean school of thought merely saw the world as particles in chaotic motion. Robert Boyle adopted this particle view in early chemistry with his *mechanical philosophy*, imagining atoms moving around and reacting with each other in a regular manner, as designed by God, not in a chaotic way.

All three of these perspectives provided helpful models and approaches for studying nature, and all, to a large degree, can be reconciled with the Christian position that a transcendent God created and controls the universe. But each has a weakness: Aristotle's works integrated with Christianity by Thomas Aquinas developed into a system of thought that placed too much emphasis on philosophy, preferring to explain events by arguing from logical first principles to deduce their cause and purpose, instead of looking for patterns through direct observation. Galileo saw himself as recovering Aristotle's own original emphasis on observations, while distancing himself from the pursuit of final causes. Neoplatonists, on the other hand, tended to explain nature's patterns as "active principles" that existed almost independently from God, or that God himself was subject to. Thus, in this perspective, God's freedom to create as he wished and his sovereignty over the universe was weakened. Finally, the mechanical philosophy focused on the interaction of particles and ran the danger of not needing God at all, for if God had set up a universe full of atoms that collide by following strict natural laws, there wasn't much left for him to do. To mechanists, something like deism or determinism started to seem more reasonable than theism.

JOHANNES KEPLER: AN EXAMPLE OF INTEGRATION[6]

Historians see Johannes Kepler as one of the most fascinating scholars from this birthing period of Western science. As a brilliant mathematician who battled health issues all of his life, his devotion to God often leapt from the page. For example, at the end of *The Harmony of the World*, he burst out:

> Purposely I break off the dream and the very vast speculation, merely crying out with the royal Psalmist: *Great is our Lord and great His virtue and of His wisdom there is no number: praise Him, ye heavens, praise Him, ye sun, moon, and planets, use every sense for perceiving, every tongue for declaring your Creator . . . and thou my soul, praise the Lord thy Creator, as long as I shall be: for out of Him and through Him and in Him are all things . . . for both whose whereof we are utterly ignorant and those which we know are the least part of them; because there is still more beyond. To Him be praise, honor, and glory, world without end. Amen.*[7]

Strikingly, Kepler's theological positions left him as a man without a country: from his writings it is clear that he was a devout

[6] In citing Kepler as a great example of integration, I do not mean to imply that all of his views are correct and that he is a perfect example for us to follow. Kepler, along with Bacon, Galileo, Newton, and the other examples I mention throughout this book, stand out because they made progress in their time and their insights contributed significantly to our modern understanding and practice of science. A person's theology and model of integration might differ from ours and even change over their lifetime, but this does not prevent their insights from benefiting us, both as Christians and as scientists. We can recognize centuries later that something Galileo or Kepler accepted has turned out to be incorrect, but that does not detract from the strides they made or the spirit they reflect in their work.

[7] Johannes Kepler, *Harmonies of the World*, trans. Charles Glenn Wallis (1939), http://www.sacred-texts.com/astro/how/how11.htm#fr_17 (emphasis original).

Christian, but he could not fully assent to the particular details of either the Catholic, Lutheran, or Reformed views. Since religion and politics were tightly joined during the counter-Reformation, more than once he was asked to move elsewhere because his doctrinal positions differed from the local governor's. Yet his passion for God and his zeal to find the patterns God established in the solar system drove him to spend most of his career trying to find a better picture for it than Ptolemy's classic geocentric model and Copernicus's heliocentric one (Kepler was frustrated that the Copernican model did not predict the planetary positions any better than the Ptolemaic model). After years of effort doing tedious and difficult calculations, Kepler reported that "as if awakening from sleep" he'd hit upon the idea of using elliptical orbits instead of circular or ovaloid ones. With this brilliant modification to Copernicus's heliocentric theory, Kepler was able to dispense with epicycles completely and to predict planetary positions thirty times better than before. When he published these results in *The New Astronomy*, he placed his revolutionary discoveries in a godly and humble context:

> The chief aim of all investigations of the external world should be to discover the rational order and harmony which has been imposed on it by God and which He revealed to us in the language of mathematics.[8]

Kepler clearly shows a deep integration of science and theology: he did not think mechanically about the world and isolate God in a box outside of it. Instead, he drew from Neoplatonic active principles and mechanistic thinking, and consciously integrated God. Space limits us to two examples:

> I believe that the cause behind the six-cornered shape of the snowflake is no other than the one responsible for the regular shapes and the constant numbers that appear in plants. And

[8] Cited in Morris Kline, *Mathematics: The Loss of Certainty* (New York: Oxford University Press, 1980), 31.

> since nothing in these matters takes place without supreme reason (not the one that is found by the labors of cogitation, but the one that was there from the very beginning in the plan of the creator, and has been preserved all this time through the marvelous nature of the animal faculties), I cannot believe that this ordered shape is present by chance even in the snowflake.[9]

> I consider it a right, yes a duty, to search in cautious manner for the numbers, sizes, and weights, the norms for everything He has created. For He Himself has let man take part in the knowledge of these things . . . . For these secrets are not of the kind whose research should be forbidden; rather they are set before our eyes like a mirror so that by examining them we observe to some extent the goodness and wisdom of the Creator.[10]

I can summarize no better than Kepler biographer Max Casper's reflection:

> Out of all the evidence adduced, and which we must restrain ourselves from increasing, the foundation on which Kepler's moral attitude grew—namely religion—is already recognizable. God is truth, and service to truth proceeds from Him and leads to Him. God is the beginning and the end of scientific research and striving. Therein lies the keynote of Kepler's thought, the basic motive of his purpose, the life-giving soil of his feeling.[11]

KEPLER'S COLLEAGUES

Kepler illustrates what we find in the other early scientists of this period: the expectation and excitement of seeing God's genius in nature. Even Galileo, while attacking the Aristotelian-infused dogma that stifled the direct study of nature, illustrates this viewpoint in his letter to the Grand Duchess Christina of Tuscany:

---

[9] Johannes Kepler, *The Six-Cornered Snowflake* (Philadelphia, PA: Paul Dry, 2012). Kindle Edition, Kindle locations 518–22.

[10] Johannes Kepler, *Epitome of Copernican Astronomy (1618–1621)*, cited in Casper, *Kepler,* 381.

[11] Casper, *Kepler*, 374. Casper's final chapter in *Kepler*, "Review and Evaluation," is an excellent summary of Kepler's thinking, compiled by the recognized dean of Kepler scholars.

> The prohibition of [Copernican] science would be contrary to the Bible, which in hundreds of places teaches us how the greatness and the glory of God shine forth marvelously in all His works, and is to be read above all in the open book of the heavens. And let no one believe that the reading of the most exalted thoughts which are inscribed upon these pages is to be accomplished through merely staring up at the radiance of the stars. There are such profound secrets and such lofty conceptions that the night labors and the researches of hundreds and yet hundreds of the keenest minds, in investigations extending over thousands of years would not penetrate them.[12]

A holistic and integrative spirit infused the scientific work of this period: Copernicus, Galileo, Kepler, and others sought patterns and mechanisms while realizing that God had orchestrated them. Excitement abounded as the improving quality of astronomical instruments and measurements were just making discernible the mathematical patterns of God's universe. Historians of science today concur that these early scientists' praise to God was not mere lip service to the cultural conventions of the times: these men were genuinely praising God for what they saw. Science, the study of nature, was truly acting as a handmaiden to draw people to worship God.

---

[12] As paraphrased by William H. Hobbs, "The Making of Scientific Theories" (address of the president of the Michigan Academy of Science at the Annual Meeting, Ann Arbor, MI, March 28, 1917), in *Science* 45, no. 1167 (May 11, 1917): 443.

THE RIFT OF THE ENLIGHTENMENT

The Notion of the World's being a great Machine, going on without the Interposition of God, as a Clock continues to go without the Assistance of a Clockmaker; is the Notion of Materialism and Fate, and tends, (under pretense of making God a Supra-mundane Intelligence,) to exclude Providence and God's Government in reality out of the World.

Samuel Clarke, writing on behalf of Isaac Newton[1]

A MECHANICAL APPROACH TO NATURE

Not everyone was happy with such an integrated picture of God and nature. Some scholars, perhaps fueled by the corruption in the church and nobility which was especially evident in France during the seventeenth and eighteenth centuries, preferred to use the regular patterns of nature as an opportunity to dismiss God entirely. They found their champion in Isaac Newton, despite his clear intentions otherwise.

Starting with Galileo's mechanical studies of motion, which Galileo published after being banned from doing further work in astronomy, Newton built a careful mathematical model for mechanical motion in his *Philosophiæ Naturalis Principia Mathematica* (*Mathematical Principles of Natural Philosophy*), first published in 1687. In his *Principia*, which is still considered by many to be the most important book in the history of science, Newton presented his three laws of motion and the universal law of gravity, which he

[1] Edward B. Davis, "Newton's Rejection of the 'Newtonian World View': The Role of Divine Will in Newton's *Natural Philosophy*," *Science and Christian Belief* 3, no. 2 (1991):103–17.

used to verify Kepler's empirical laws of planetary motion. The blockbuster success of Newton's genius at explaining "how the heavens go," by taking simple principles that could be observed and studied on earth and extending their application throughout the universe, cemented the loss of confidence in Aristotle's physics. What previously was imagined as a universe governed by a "world soul" with the non-mathematical desires of a living being was now shown to follow simple and elegant mathematical equations. Aristotle's picture of the universe disintegrated: the earth was not the sump or garbage pit down to which all the refuse of the heavens fell,[2] and the heavens above could no longer be imagined as unchanging and composed of an ethereal material unlike anything on earth. Newton showed that Copernicus's revolutionary model was correct: the earth was actually one of the heavenly bodies, not a pit, and the mathematical laws of physics that Newton had discovered on earth also governed the other planets and the rest of the cosmos. Aristotle missed it: mathematics had a power which he had not anticipated.

With Aristotle's worldview discredited, what other conceptual framework could natural philosophers turn to? Newton himself joined many others in seeing God as sovereign over a completely passive physical world where God is the active principle who draws particles together, and which we observe, for example, as the force of gravity. This "sovereign-passive" relationship closely paralleled the Reformation theology of Martin Luther and John Calvin, who saw God's salvation as a sovereign gift to sinners, who cannot work or do anything to merit it, but can only receive it passively as a gift. Their concept of "passive righteousness" stood in sharp contrast to the Roman Catholic Church's idea of participatory grace, where

[2] The popular story that Copernicus's heliocentric view of the solar system removed humans from their pedestal as the central beings of the universe is simply a modern myth and grossly misrepresents the importance of the earth in the Aristotelian model. For more details see Dennis Danielson's excellent chapter in *Galileo Goes to Jail and Other Myths about Science and Religion*, ed. Ronald L. Numbers (Cambridge, MA: Harvard University Press, 2009).

believers earned favor from God through their involvement in various church rites and ceremonies.

Although Newton could show that the force of gravity was not the result of particles colliding with each other, he recognized that he could not prove his idea that God was the active force in moving passive matter, because at our level we can only see principles and equations that correlate matter's behavior. Thus when critics attacked his *Principia*, charging him with not adequately explaining what gravity really was, he replied that he "feigned no hypothesis" about the mechanical cause of gravity: he only intended to characterize its operation. The contrast here is interesting: Newton's laws of motion focus on the *efficient* or *primary* mechanical causes for particle interactions instead of speculating about *final* or *purposeful* causes (recall that Aristotle had argued that the final cause was most important, because once the system's teleology was known, the other causes logically fell into place). But to Newton, gravity required the operation of a nonmechanical active principle. Thus Newton found fault with both the organic Aristotelian and the mechanical views of nature. Unfortunately, over time the success of his mechanical laws of motion became the wedge used by others to eliminate God and any notion of purpose from science.

THE MECHANICAL RATCHET

As mathematics and measurements improved, the increasingly apparent regular operations of the physical world caused skeptics to question the idea that God himself was responsible for moving the particles in exactly the same way every time. Many theistic scholars addressed this problem by embedding the active principles within nature as internal forces, so that God's direct action to animate it was not necessary. A favorite analogy was to imagine the universe as a giant mechanical clock with God as the clockmaker. Many historians find it striking that Christian thinkers helped lay the groundwork for the shift from a theistic to a deistic picture of the universe as they debated God's exact role in nature. Gottfried Leibniz, for example, argued that God could only have created the best of all possible worlds, and therefore he had made a world machine which operated without any need for his intervention or guidance, apart from miracles of grace. Robert Boyle, often called the father of chemistry, was a very devout Anglican, yet he helped coin the term "mechanical philosophy" and argued for a clockwork universe as a means to establish God's sovereignty. Even René Descartes, who introduced skepticism to philosophy and argued for an entirely mechanical universe, was seeking to defend the spiritual reality of the human soul. The intent here of Boyle, Descartes, and others was to counter the prevailing view that the world was organic or magical in nature; instead they insisted that God established it to run according to fixed laws. As mentioned, Newton resisted this reductionist way of thinking about the world as merely particles in motion, arguing that the force of gravity could not be derived from particles colliding with each other, and preferring instead the Neoplatonist image of gravity as an immaterial active principle.

It was the skeptics, however, who leveraged the advances in mathematics to argue that perhaps God was not necessary at all. What came to be called "Newtonian physics" reflected the mathematics, but not the conceptual framework, of Newton himself. What Newton saw as a theistic universe, where God guided and influenced the physical world in both regular and special ways, became to many of Newton's followers a deistic clockwork universe, which God had created, but in which he did not interfere. Moreover, David Hume, Voltaire, and Pierre-Simon Laplace influenced the general public as much as the scientific community with their popular-level arguments that a mathematical, mechanical, and material approach to the world was so highly successful at explaining everything that the God hypothesis was no longer necessary. Instead of a sovereign personal God overseeing, guiding, and caring for his handiwork, the clockwork universe ran because of inflexible and impersonal laws.

PARADISE LOST

With this shift of perspective, the Christian tradition that God is (or could be) active in nature was effectively lost in the sciences. When science was starting, the regular patterns in nature were seen as evidence of his handiwork and control over his creation. As these patterns were explained as the operation of God's fixed laws, laws which were inherent in the material world and not externally directed by God, scientists tended to shift from a theistic to a deistic approach. Yes, God created the world so that it runs in these incredible ways, but he doesn't actively influence it today. There was still talk of God's transcendence, but his immanence was ignored, giving an unbalanced perspective. From here it was a simple process to push God even further into the distance, leaving agnosticism or even atheism to fill the vacuum. As the sciences became increasingly secular in their perspective, however, they continued to borrow the fundamental Christian assumptions about the rationality and regularity of nature. As we will see later, this raises some critical difficulties for the logic and practice of modern science because it is hard for a non-theist to justify these assumptions with more than an ad hoc "it's worked so far" basis.

Of course, this loss of a Christian perspective was gradual and there are many notable scientists in the nineteenth century who carried an integrative mind-set in their work[3]: Michael Faraday, considered one of the greatest experimental scientists of all time for his pioneering work in electricity, magnetism, and chemistry, sought to uncover the design in God's book of nature by seeking the laws he had impressed on nature. In his biography detailing both the scientific and devout Christian life of Faraday, Geoffrey

[3] Space limits us to mentioning only two by name. An accessible resource for others is Dan Graves, *Scientists of Faith: 48 Biographies of Historic Scientists and Their Christian Faith* (Grand Rapids, MI: Kregel, 1996).

Cantor notes, "The consistency and simplicity of nature were not only *conclusions* that Faraday drew from his scientific work but they were also metaphysical *presuppositions* that directed his research."[4] Also James Clerk Maxwell, ranked in physics with Isaac Newton and Albert Einstein for his work on electromagnetic waves and light, took an integrative perspective in science because of his deep Christian convictions. In an article published in the journal *Nature*, Maxwell argued for design based on the uniformity of molecules, writing,

> None of the processes of Nature, since the time when Nature began, have produced the slightest difference in the properties of any molecule. We are therefore unable to ascribe either the existence of the molecules or the identity of their properties to the operation of any of the causes which we call natural. On the other hand, the exact equality of each molecule to all the others of the same kind gives it, as Sir John Herschel has well said, the essential character of a manufactured article, and precludes the idea of its being eternal and self-existent. Thus we have been led, along a strictly scientific path, very near to the point at which Science must stop.[5]

THE SECULARIZATION OF SCIENCE

Unfortunately, what had begun as skepticism about the reliability of Aristotle grew to become a resistance to any ancient authority. Scholars in the Enlightenment movement argued that since Aristotle was not reliable, other ancient sources were suspect as well. Therefore, truth could only be established through contemporary human *reason and experience*. This rationalist approach seemed to work in most cases: through the careful application of logic, mathematics, and experimental methods, reason and experience were bearing fruit, discovering mechanical, regular laws that appeared to leave no room for, or even need for, God.

[4] Geoffrey Cantor, *Michael Faraday: Sandemanian and Scientist* (London: MacMillan, 1991).
[5] James Clerk Maxwell, "Molecules," *Nature* 8, no. 204 (September 25, 1873), 437–42.

This secularization movement continued into the nineteenth century with positivism and scientism, philosophies which argued that only *scientific* knowledge is true and authoritative. As these views influenced those writing science and history books, it became easy to ignore the theological content and context of earlier scientists, to dismiss their religious remarks as cultural accommodations to their times, or worse, to see the emergence of modern science as a victory over ancient superstition.

NATURALISM COMES TO BIOLOGY

While naturalism came into physics in part because of the predictive power of mathematics, the arguments bringing naturalism into biology a century later took a different approach. Although mechanical and machine analogies had been applied to animals before, organisms' levels of complexity were far greater than the simple gears, levers, and springs which humans made, so "Nature" was obviously a far better engineer than we were. But there was something about biology which did not square with the nineteenth century Victorian notion of a good, benevolent God who created this world for our happiness: there was too much pain, suffering, and inefficiency in it. This could be explained as the result of the fall, of course, but it grated on Charles Darwin, especially after the death of his favorite daughter, Anne, when she was ten years old.[6] Thus Darwin insisted that his theory that gradual, slight changes occurring in plants and animals over many generations gave rise to the complexity they have today, must have occurred without the need for a guiding hand. As Asa Gray and some other contemporaries quickly pointed out, there is nothing in Darwin's mechanisms of gradual adaptation over time which rules out the possibility that God is directing the changes, yet such a guided interpretation of evolution was rejected by Darwin, T. H. Huxley, and other spokes-

[6] Adrian Desmond and James Moore, *Darwin: The Life of a Tormented Evolutionist* (W. W. Norton, 1991), 387.

men for the new theory. *The whole point of Darwin's approach was to explain how life got here without invoking God.* This still remains an issue today, as secular scientists commonly regard "theistic evolution" as an oxymoron, since they define "evolution" as an unguided, purposeless, and undirected process. Some theists prefer this unguided interpretation too, because it allows them to distance God from the pain and suffering in the world.[7]

As much as Darwin wanted to tell a creation story that did not require God at all, Darwin had to concern himself with data that showed the sudden appearance of new and different life forms in the fossil record, facts which favored the standard creationist view for the origin of life and its diversity. His tactics to dismiss a creationist interpretation involved brilliant rhetoric:

> If it could be demonstrated that any complex organ existed, which could not possibly have been formed by numerous, successive, slight modifications, my theory would absolutely break down. But I can find out no such case.[8]

Despite large and systematic gaps in the fossil record, such as the Cambrian explosion, Darwin claimed to be aware of "no such case" in his day. His reasoning here is crucial: note the key phrase, "could not possibly." Philosophers refer to this type of statement as a *universal negative*, and they often leave the room when such arguments come up, because there simply is no logical way to defeat them. To rephrase Darwin's argument more clearly, he is asserting that his gradual, unguided evolution model is correct unless

[7] For example, Francisco J. Ayala notes, "Later, when I was studying theology in Salamanca, Darwin was a much-welcomed friend. The theory of evolution provided the solution to the remaining component of the problem of evil. As floods and drought were a necessary consequence of the fabric of the physical world, predators and parasites, dysfunctions and diseases were a consequence of the evolution of life. They were *not* a result of deficient or malevolent design: the features of organisms were not *designed* by the creator," *Darwin's Gift to Science and Religion* (Washington, DC: Joseph Henry, 2007), 4–5, italics in original. See also Cornelius G. Hunter, *Darwin's God: Evolution and the Problem of Evil* (Grand Rapids, MI: Baker, 2001).

[8] Charles Darwin, *On the Origin of Species by Natural Selection, or, the Preservation of Favoured Races in the Struggle for Life*, Kindle edition, chap. 6, based on the first edition.

someone can prove that it is *absolutely and totally impossible* for a complex organ to form by a gradual process. *There is no need for Darwin to prove that it actually did form by a gradual process*; he can simply assert that he is correct because it is *logically possible that it could have* evolved by numerous, successive, slight modifications. Darwin shifted the burden of proof to the skeptic and raised it to an impossible level.

I must admit that I was puzzled when I first encountered Darwin's argument, since it does not reflect standard scientific practice. To illustrate, let's imagine that Renoir's *Girls at the Piano* appears on the wall of my office one afternoon. Colleagues ask me where this masterpiece came from, and I say that I've been practicing painting for some time and finally decided to hang one of my pieces here in my office. Suspicious, they investigate my claims and find out that the original has been reported stolen from the Musée d'Orsay in Paris, that I don't have an art studio at home, and that the paint on the canvas uses chemicals which have not been commercially available for decades. I protest, however, that they are not able to prove that I *could not possibly* have painted it, so therefore they must accept the fact that I did.

Obviously, law courts don't work this way, nor should science. The standard practice in both areas follows what philosophers call "inference to the best explanation," beyond a reasonable doubt. One seeks the most likely and reasonable explanation among a range of possibilities, instead of asserting that one's own proposal is correct unless others can absolutely prove it to be impossible.

WHY WAS DARWIN SO PERSUASIVE?

So why did Darwin's clever argument carry the day in biology, and why is it still defended? There were several reasons: first, it was the best naturalistic explanation that excludes God, helping free the mind from cosmic authority figures. Second, it seemingly avoids the problem of evil, since "evolution did it" instead of God, as we already noted. Third, it was politically advantageous. T. H. Huxley and others seized upon *On the Origin of Species* for this last reason, because Darwin told a secular creation story which allowed them to push religious authority out of the public square and to replace it with the authority of science. Herbert Spencer favored a political interpretation of evolution and advocated a form of "social Darwinism," a broad concept which quickly degenerated to justify war, racism, and eugenics as part of the human struggle for the survival of the fittest. While the dark side of these political versions of Darwinism is generally recognized today, the cultural prestige that evolution still enjoys harkens back to Darwin's success in explaining how we got here without the need for God. After all, if the physical universe is just a mechanical clock and all living things gradually improve over time on their own, then what need is there for a Creator? It seems that Darwin won with rhetoric and by having a naturalistic answer that many people wanted, and still want, to hear.

DARWIN AND THE PROBLEM OF EVIL

Let's look more into that concern about the problem of evil: while a detailed discussion of this topic would require a book of its own, it is worth mentioning briefly because it touches on biology. Clearly, we live in a world filled with pain and suffering, and it is virtually impossible not to be touched by them and not to have a sense that "this ought not to be." The Christian answer to this universal problem is profound: God also agrees that evil ought not to be, and, in fact, this creation is a battleground between good and evil as God fights to destroy evil once and for all. In particular, God is trying to rescue *us* from evil after we freely chose to rebel against him and to be evil ourselves! So why do pain and suffering happen? They are the consequence of our rebellion.

As Jesus comments in Luke 13:1–9, one reason bad things occur is to "fertilize" us to repent and turn back to God. Thus, instead of complaining about pain, we should be astonished when anything good happens to us, because goodness shows the grace, patience, and mercy of God. Amazingly, the apostle Paul notes in Romans 8 that those who repent and turn to love God have the promise that everything will somehow work out for good in the end. And here's what I think is the most incredible part of the Christian answer to the problem of evil: God in Jesus suffered, himself, on the cross—both physically and spiritually—in order to rescue us, with pain levels worse (given his divine nature) than any other person or creature ever has or ever will experience. While we don't have an answer for any particular ache, pain, or atrocity, the Christian perspective faces evil's reality squarely and promises a rescue—a rescue made at tremendous cost to God himself.

As already mentioned, some theists use the evolutionary process as a means to distance God from the pain, suffering, and inefficiency we see in the world. It's a simple argument: God set up the evolutionary process, but since he didn't guide it, he isn't responsible for the way things turned out. But wait: Just why isn't God responsible for the way things turned out? If he created it, couldn't he have done it some better way? After all, we hold a dog owner responsible if his pet injures someone, especially if he does nothing to curb the dog's aggressive behavior after repeated warnings.

Arguing that the Creator isn't responsible for the outcomes of the process that he supposedly established is not satisfying, but more importantly, it isn't the biblical picture either. Granted, the biblical picture is a bit shocking: in Exodus 4:11, after Moses complains that he can't go to Pharaoh because he doesn't speak very well, God says to him, "Who has made man's mouth? Who makes him mute, or deaf, or seeing, or blind? Is it not I, the LORD?" Stunningly, God lets the buck stop with him when it comes to the cause of disabilities. But why does God allow them, if he could prevent them? There's no simple answer to this, but somehow pain and suffering are part of God's plan. Without being evil himself, God permits ugly things to happen to accomplish great goals, such as developing our character through trials. How the details of all this work out has kept theologians scratching their heads for centuries, but one thing is clear: we don't need to shelter God from this issue. He has promised that somehow all of this evil is going to work out for good, and since we are humans who can only peer through a glass darkly, we may have to leave it at that, recognizing that he paid a great price so that there will be a happy ending.

SUMMARY

The natural sciences, which began as a puzzle-solving quest to think God's thoughts after him, became too enamored with the universe's mechanical and mathematical regularities, which made some wonder how God could actively guide what appeared to be a rigid and closed system. Thanks to nudges from secularizing cultural pressures, which argued that truth could be established only on the basis of human reason and experience, God's involvement with the world tended to be reduced to that of a transcendent, deistic clockmaker. From there it was an easy step to discount him entirely, leaving an eternal, deterministic, fatalistic, and impersonal universe in which life somehow had gradually evolved through random processes to the point that one species, at least, could wonder why it was here. But, as we will see, perhaps these nudges and steps were taken too quickly.

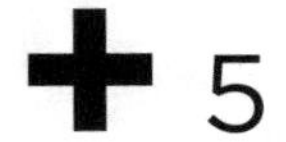

THE WRONG ROAD

SCIENCE AS METHODOLOGICAL NATURALISM

Moreover, that materialism is absolute, for we cannot allow a Divine Foot in the door. To appeal to an omnipotent deity is to allow that at any moment the regularities of nature may be ruptured, that miracles may happen.

Richard Lewontin[1]

Today, the Enlightenment's appeal to "reason and experience" goes by the term *methodological naturalism*. Its central idea is that scientific answers can make reference only to natural, physical causes for any effects that we observe. Natural forces like the mixing of warm and cool air masses are thus acceptable explanations for rain, but invoking God as a cause of the storm would not be.

Methodological naturalism is often touted as religiously neutral; that is, a person from any religious tradition—or from none—can practice science equally well under its guidance. This seems logical since in many situations there is no particular reason to evoke a higher power as the cause for an event, because the ordinary working of physical laws and "chance"[2] are sufficient.

Yet, as innocent and universal as the methodological-naturalism approach to science sounds, it has one serious problem: it is the default

[1] Richard Lewontin, "Billions and Billions of Demons," review of *The Demon-Haunted World: Science as a Candle in the Dark*, by Carl Sagan, *The New York Review of Books*, January 9, 1997, http://www.nybooks.com/articles/archives/1997/jan/09/billions-and-billions-of-demons/?insrc=toc.
[2] "Chance" is a difficult term to define, but at least it covers our ignorance of the precise initial conditions behind an event.

philosophical position of the atheist or deist, who believes either that God doesn't exist or that he does not influence the universe in any way today. The practitioner of any other religion is told that he must adopt this alien philosophy in order to do good science.

How is this a problem? Well, let's imagine that you're driving a bus, and you leave the main terminal with three passengers. At the first stop you pick up two, at the next six, but you let off one. At the third stop you pick up four and let off two, at the fourth stop you let off one and pick up three, at the fifth stop you pick up seven, at the sixth stop you let off five and pick up one, and at the seventh stop you let off three and pick up two. Now, as you approach the eighth stop, what is the name of the bus driver?

Like this classic joke, methodological naturalism can get you so focused on the details of physical cause and effect that you forget who you are or, rather, that God is involved in *everything*. As we will discuss further, a Christian in science will certainly prefer to evoke physical causes for regular physical events; but to extend that preference to make it an automatic exclusion of any possibility that nonphysical causes ever act in the world seems to be extreme. Is it necessary for a theist to adopt a "pragmatic naturalism" in order to do good science? This naturally leads us to the question, what is "good science"?

DEFINING "SCIENCE"

I have avoided defining *science* up to this point because, as with many concepts, everyone has a general sense of what it means, but it is difficult to define precisely. In fact, many philosophers have given up on seeking a simple definition of *science*, since its methodology is not unique and it is impossible to isolate it from philosophical and theological considerations.[3] But a concise, rough definition is the following: *Science is the study of the physical world by using*

[3] See for example, J. P. Moreland, *Christianity and the Nature of Science: A Philosophical Investigation*, 2nd ed. (Grand Rapids, MI: Baker, 1999).

empirical observations to infer patterns and relationships that have predictive value and can be applied for useful ends. There's a lot more that could be added and qualified here,[4] but our focus for the moment is on the phrase "patterns and relationships." Note that science practitioners must infer *culturally acceptable* patterns and relationships. Thus the ancient Babylonians were "scientific" in taking careful and systematic measurements of the positions of the planets (gods) in the sky, in order to infer the gods' disposition and to guide the king's activities. Modern science, through the constraint of methodological naturalism, insists that the only culturally acceptable "scientific" patterns and relationships are physical interactions. This is fine for atheists and deists, but troublesome for theists, especially in questions involving origins, where a theist might expect to see God's hand.

THE GOAL OF SCIENCE

To see more clearly where the trouble lies, we need to think more about the goal of science. For almost everyone today, science is more than a search for useful patterns and relationships; *it is the search for truth.* Scientists and the media commonly portray science in this light. Philosopher Daniel Dennett, in a lecture at Oxford University, commented:

> And year after year, the elite in every nation in the world send their children to our universities for their educations. They know, perhaps better than we ourselves appreciate, that the science and technology of truth-seeking is our most valuable export.[5]

Of course, since Dennett is an atheist, he sees no difference between material or natural causes and truth. But is it his religious

[4] For example, "useful ends" can refer to more than technology, such as a better theoretical understanding of the fundamental nature and structure of matter, which is sought in the field of high-energy particle physics. Classically, science is one of the liberal arts because it broadens and deepens our understanding of and appreciation for the world.

[5] Daniel C. Dennett, "Faith in the Truth" (Amnesty Lecture, Oxford, February 18, 1997), http://ase.tufts.edu/cogstud/papers/faithint.htm.

commitments that are making this connection, or are material or natural causes fundamental to the definition of "science?"

Richard Lewontin, a geneticist at Harvard, raised this point in the context of Darwinism:

> We take the side of science in spite of the patent absurdity of some of its constructs, . . . in spite of the tolerance of the scientific community for unsubstantiated just-so stories, because we have a prior commitment, a commitment to materialism. It is not that the methods and institutions of science somehow compel us to accept a material explanation of the phenomenal world, but, on the contrary, that we are forced by our a priori adherence to material causes to create an apparatus of investigation and a set of concepts that produce material explanations, no matter how counter-intuitive, no matter how mystifying to the uninitiated.[6]

Notice that Lewontin is not arguing that science has a *preference* for material or natural causes, but that the goal of science is to give naturalistic answers *no matter what*.

Strangely enough, I first became aware of this naturalistic-no-matter-what goal of science in a Bible class, when the professor remarked that as scholars we could not take the miracle accounts in the Bible to be historical events, because "that would not be viewing the text scientifically." Now I'd gotten my PhD in Physics by swearing to tell the whole truth and nothing but the truth in my lab research, and here I was, pursuing a PhD in Ancient Near Eastern Studies and being told that in order to be "scientific," I could not even consider the possibility that these miracles had actually happened. Why must I exclude a possible hypothesis? Because "scientific" is equated with "naturalistic."

It turns out that I had somehow missed what some think is one of the basic rules of science. Perhaps this happened because I had worked in membrane biophysics, where material causes were

---

[6] Lewontin, "Billions and Billions of Demons."

adequate to explain the immediate physical effects, but now that I was working on broader theological, historical, and origin questions, I had stumbled into this invisible wall.

But I'm not sure that science can honestly say that it has both of these goals: if science is the pursuit of truth, then it should go wherever the evidence leads, without concern for culturally acceptable answers, be they materialistic or otherwise. If science provides materialistic answers no matter what, then it is philosophically bound to produce materialistic answers *that may not be true*. I find this disturbing, because if a materialistic explanation must always be preferred in order to be "scientific," then an intelligent cause could never be considered even if it were true.

An example from theology illustrates this problem: the New Testament reports that Jesus fed thousands of people on more than one occasion by multiplying a few fish and loaves of bread. If we are seeking truth, then the reports of a miracle by credible eyewitnesses should be given serious consideration. If we are committed to finding the best materialistic answers no matter what, however, then we must explain away these reports, perhaps by saying that the crowds were shamed into sharing food they had hidden, because of the generosity of the boy who gave Jesus all he had. Such an answer is considered "scientific" because it is a naturalistic way to *explain away* the reports, not because that is what actually happened.

Strikingly, many of the founders of modern science would be as shocked at this naturalistic definition of "science" as I am. Max Casper, Kepler's premier biographer, summarizes Kepler's keynote thoughts as "God is truth, and service to truth proceeds from Him and leads to Him. God is the beginning and the end of scientific research and striving."[7]

If God started out being the beginning and end of scientific research, why does Lewontin insist that science must hold such a firm commitment to materialism today? He continues:

[7] Max Casper, *Kepler*, trans. and ed. C. Doris Hellman (New York: Dover, 1993), 374.

> Moreover, that materialism is absolute, for we cannot allow a Divine Foot in the door. To appeal to an omnipotent deity is to allow that at any moment the regularities of nature may be ruptured, that miracles may happen.[8]

Unfortunately, Lewontin has the history of science backwards: Copernicus, Galileo, Kepler, Newton, Boyle, and many other early modern scientists justified their search *for* the regularities of nature precisely because there *was* an omnipotent God behind it. Oddly, it is the materialist who has no good reason for why the sun should come up tomorrow morning, other than the extrapolation based on what it did yesterday. Investors are often reminded that the past performance of the stock market is no guarantee of future returns; similarly, the materialist presumption of regularity can make no guarantees about tomorrow. It has borrowed from Christian capital to presume that the laws of physics will stay the same in the future.

A DIVINE FOOT IN THE DOOR?

But what about that "Divine Foot in the door"? As we've seen, the Judeo-Christian expectation is that God normally works in regular ways through secondary causes in order to achieve his ends, so science grew and functioned very well by preferring physical causes for regular physical effects. But Christianity sees no reason to bar God from being the primary cause of things from time to time, especially in three areas: origins, history, and theology. Actually, for a naturalist these three fields can get to be rather boring, since he already knows the "correct" answer: in origins research, something like Darwinism and the multiverse must be defended, because any other answer is too theological; history is simply a long list of wars and politics, because any direction or goal for humanity is too theological; and theology itself becomes the art of explaining away

[8] Lewontin, "Billions and Billions of Demons."

any actual human contact with or evidence for God. Even atheists like Thomas Nagel are bothered by these automatic materialistic answers to such significant questions, and in a candid moment he speculated about its cause:

> My guess is that this cosmic authority problem is not a rare condition and that it is responsible for much of the scientism and reductionism of our time. One of the tendencies it supports is the ludicrous overuse of evolutionary biology to explain everything about life, including everything about the human mind.[9]

While methodological naturalism reflects a religious preference for exclusively physical causes, the Christian intellectual tradition in the sciences is a holistic one, seeing God in material causes and not being surprised if we see evidence of his hand at key points in our history and origin. Thus the Christian in science is able to pursue truth and follow the evidence wherever it leads and enjoys the freedom to think outside the naturalistic box.

COMMON GROUND

But if modern science has such a single-minded focus on materialistic explanations, is it possible for Christians to practice science today? In a word, yes.

Most scientific research and practice involves questions not related to origins or history, and thus material explanations are satisfying to a range of secular and sacred worldviews. While a theist would see God's hand guiding and sustaining the physical world, this mostly results in regular and law-like behavior that a secular person would accept at face value. Close and deep cooperation is possible because the patterns and relationships being studied fit within both perspectives.

What about the subjects of origins, history, and theology? I

[9] Thomas Nagel, *The Last Word* (New York: Oxford University Press, 1997), 131.

find that common ground is still possible, but it may depend on the perspective of the individual Christian (or non-Christian). The range of modern perspectives is well summarized by Ian Barbour, who has devoted his career to the study of science and religion. Barbour recognizes four broad approaches in topics where science and religion might overlap[10]:

1. CONFLICT

As it takes two parties to start a conflict, there are two groups here. On one side are those who believe that science is the only path to truth and that religion is simply superstition and myth; on the other are those who believe that their interpretation of the Bible is absolutely correct and that science is but a conspiracy to prove atheism. Conflict between these two radical views is inevitable, and, since it makes for a good media story, it is the most common view that the average person is aware of today. Despite this tension, two scientists from these ends of the spectrum can fruitfully work together: Kurt Wise, a young-earth creationist, completed his PhD thesis under the guidance of Stephen J. Gould, an atheist.[11] They were productive because they focused on their common interests instead of their differences.

Although it may be common today, historians of science do not see the conflict model as significant in the historical Christian intellectual tradition. Certainly there were some tension points in the past between the interpretations given to physical data and Scripture, but the common goal until recently was to seek harmony between the two, instead of demanding an either/or acceptance of one extreme or the other. It is worth noting again that what fuels the conflict model is the tension between Christianity and

[10] See Ian G. Barbour, *When Science Meets Religion: Enemies, Strangers, or Partners?* (New York: HarperOne, 2000). Barbour writes from a theologically liberal Christian perspective, but he has a broad familiarity with the critical issues.

[11] The key factors are to be polite and respectful of each other, and to focus on common ground. As another example, Will Provine and Phil Johnson were polar opposites in their science and religion perspectives, yet as they debated each other in the 1990s, they became good friends.

naturalism, two different religious positions, not Christianity and the physical data; moreover, naturalism was not a significant player in Western Europe until the latter half of the eighteenth century.

2. INDEPENDENCE

People with this independance perspective see science and religion as very different disciplines with distinct interests and no significant overlap. "Science tells us how, religion tells us why" is a simple summary. Science focuses on objective data, while religion deals with subjective meanings. There certainly is value in this approach, because it eliminates controversy, but its terms are harsh: the two disciplines are isolated and barred from any meaningful interaction. Science and theology are simply two different ways of knowing, so as they look at the same things from their different perspectives, they may draw different conclusions. Both conclusions can be true within the limits of each perspective—if science recognizes that its focus on material causes does not allow it to address metaphysical questions, and if theology recognizes that its focus on meaning and significance does not speak to material cause and effect. A scientist who is a Christian and who holds an independence view would find no problem with any materialist models for origins because he does not expect the Bible to have anything to say about material origins. Thus he has no problem working with secular scientists because he also assumes a materialistic perspective when looking at the world; God works in the background as the sustainer and Creator, but is not discernible by science. Theology is thus an additional layer that his colleagues may not have.

Unfortunately, the independence approach has the danger of hiding a darker goal than isolation: it can be used as an effective but less hostile way to dismiss religion than the conflict model. After all, if science has the facts and religion has only subjective values, then public policy and education should be determined by facts, and personal values should be left at home. Because of these possible abuses, we will review this perspective in more depth later.

3. DIALOGUE

Scholars specializing in the history and philosophy of science often prefer this dialogue approach. It recognizes that science and theology do overlap somewhat, especially since the Christian culture in which science developed shaped its assumptions and goals. Friendly discussions on contemporary topics of common interest (such as bioethics) are possible, without one side challenging or threatening the other's interpretations. The areas of common interest and discussion, however, are limited—perhaps too much so—to metaphysics, history, and ethics. A Christian working in the sciences who is well-versed in the dialogue perspective might recognize the weaknesses and limitations of a materialistic approach to origin questions and could bracket them with a that's-the-best-science-can-offer disclaimer.

4. INTEGRATION

Some scholars call this integration perspective the "classical" approach because it aims for a holistic view of God and nature, as is so often apparent with the early scientists. Science, like archaeology, is able to inform theology in areas of overlap, just as theology is able to inform scientific interpretations in areas of overlap (such as origins and history). Harmony is sought between the two disciplines as much as possible, and any differences which remain suggest that an *interpretation* of the data and/or Scripture is at fault. A Christian with this perspective is comfortable with materialistic explanations until they begin to take on metaphysical or religious tones or until they appear to clash with a clear scriptural interpretation, in which case harmony is sought by exploring other interpretations or by collecting additional data to see if that resolves the apparent problem.

CAN WE LIMIT SCIENCE?

Some of my colleagues think that it would be best for science to limit itself strictly to physical cause and effect, and avoid concerning itself with broader theological issues. This would require an enormous shift in today's culture, as many scientists have become comfortable with making theological pronouncements and in claiming the high ground of having facts and truth relative to other disciplines. Are scientists actually capable of sticking only to material causes and leaving the obvious metaphysical questions dangling? I imagine this is possible: science could impose boundaries on itself and say, "This is the best physical cause that we can determine, but of course there may be immaterial forces at work here that could offer a much better explanation for these phenomena than we can." Perhaps such a constant reminder of a limit to science might work, but I don't see many scientists rushing to add a footnote like this to every page of their books or journals. Why not? Because the metaphysical questions that naturally arise when we study certain aspects of the world around us are too obvious, interesting, and compelling to ignore or to pass over to the experts in another discipline.

Rather than limit science, my conclusion from studying the history of science is that it would be more appropriate to recognize that physical data are always interpreted through metaphysical lenses. A healthy and fruitful practice of science would allow scientists to offer competing interpretations of the data openly through a variety of lenses. Although this diversity is rarely glimpsed in science textbooks, scientific knowledge, like all human endeavors, progresses best when a variety of interpretations can be aired, argued, and compared. Certainly in some areas a range of metaphysical perspectives is as important to consider as the range of mathematical or physical models. For example, the atheist philosopher Thomas Nagel recognizes the need for more metaphysical openness when it comes to discussions of evolution and intelligent design (ID):

> The claim that ID is bad science or dead science may depend, almost as much as the claim that it is not science, on the assumption that divine intervention in the natural order is not a serious possibility. That is not a scientific belief but a belief about a religious question: it amounts to the assumption that either there is no god, or if there is, he certainly does not intervene in the natural order to guide the world in certain directions.[12]

In this same article, Nagel reaches the following conclusion:

> The consequence of all this for public education is that both the inclusion of some mention of ID in a biology class and its exclusion would seem to depend on religious assumptions. Either divine intervention is ruled out in advance or it is not. If it is, ID can be disregarded. If it is not, evidence for ID can be considered. Yet both are clearly assumptions of a religious nature. Public schools in the United States may not teach atheism or deism any more than they may teach Christianity, so how can it be all right to teach scientific theories whose empirical confirmation depends on the assumption of one range of these views while it is impermissible to discuss the implications of alternative views on the same question?[13]

CAN WE LIMIT METHODOLOGICAL NATURALISM?

I fear that there may be a personal impact from adopting methodological naturalism as a Christian, working scientist. If you have to leave God at the door of the lab when you go to work, and you work many hours a day without thinking about him, then it becomes easy to forget to pick up God at the door again when you leave to go home. Imagine the protests that atheists would mount if they had to assume and act as if God existed while doing their professional work, but were told that they were free to believe what they pleased elsewhere.

---

[12] Thomas Nagel, "Public Education and Intelligent Design," *Philosophy & Public Affairs* 36, no. 2 (2008): 198.
[13] Ibid., 200.

Of course, consistent Christian living is difficult, and one of the criticisms we often face is hypocrisy—that we say we believe one thing but do another. Yet the mainstream approach to science today encourages exactly such a disconnected behavior for Christians. This is why I find Kepler so refreshing when he pauses in the middle of a technical treatise to praise God: in Kepler's mind there is no separation of God from science.

But there is a deeper problem than mere hypocrisy here. Methodological naturalism encourages the tendency to think of the world in an either/or dichotomy: there are natural events, and then there are supernatural events. A Christian believes that God set up the natural events, of course, but he could then think that God doesn't concern himself with them or guide them. Thus the only time that he really sees God at work is when something spectacular like a miracle occurs. Unfortunately, since miracles are rare, God doesn't seem to be doing much.

This natural-supernatural dichotomy surfaces when people ask questions like, "Does it pay to pray?" The presumption is that unless we're going to ask God to do a miracle, God doesn't influence or guide the workings of the world in any way. When my son was six years old, he asked me this question after we had given thanks for dinner one evening: "Dad, why do we thank God for the food? Mom's the one who prepared it." He did have a point—it's not as if we sat down at an empty table, closed our eyes and prayed, and then "poof," a banquet appeared after I said, "Amen." So this started a discussion about how God had blessed me with a job, so we had the money to buy food; how God had blessed the farmers, so their crops and animals grew; the long supply chain to the grocery store, so that there was food to buy; and how God had given Mom the health and skill to prepare it. If we close our eyes to seeing God at work around us in his regular ways, because that's what we must do in order to be "scientific" or because we pick up this blindness from our secular culture, it's not hard to become deistic or even agnostic "Christians." Often this drift is gradual. William Sumner, the Yale sociologist who had trained and served as an Episcopal priest in his early years, famously remarked, "I never consciously gave up a religious belief. It was as if I had put my beliefs into a drawer, and when I opened it, there was nothing there at all."[14]

Such a dangerous drift toward a functional naturalism is inherent in public education today. One evening I was helping my son with his ninth-grade biology homework, which was covering Darwinian evolution. At one point he asked, "Why is this stuff so different than what we learned in Sunday school? Where is God in this?" After a quick prayer for wisdom, I answered that since he was in a public school, they couldn't talk about God, because that was religious. To this he said, "Well, if we don't need God here, why do we need him anywhere?" After another, longer, prayer for wisdom I answered, "Well, they're just trying to tell the best naturalistic story they can by leaving God out of it, and that sure leaves a lot of holes, doesn't it?" To which he said, "Oh wow, you're right, they can't explain . . ." and he went on for a couple of minutes listing the problems with Darwinism.

But Lee Strobel wasn't so lucky. In *The Case for a Creator* he relates how a similar thought struck him in his high school biology class: "If the origin of life can be explained solely through natural processes, then God was out of a job."[15] When he got no reasonable answers to his questions, he proceeded down a path of atheism for many years.

As we saw before, atheist philosophers like Thomas Nagel are concerned about the religious bias of assuming methodological naturalism, especially when teaching origins in public school science classes. The atheist Michael Ruse has made similar observations on many occasions, noting that the real cultural battle is not

[14]Quoted in George M. Marsden, "God and Man at Yale (1880)," *First Things* 42 (April 1994): 39–42.
[15]Lee Strobel, *The Case for a Creator: A Journalist Investigates Scientific Evidence That Points toward God* (Grand Rapids, MI: Zondervan, 2005), 19.

between Christianity and science, but Christianity and naturalism, two very different religious views about the world. This is especially evident regarding evolution, where Ruse notes that Darwinism can function at the popular level like a secular religion.[16]

CAN SCIENCE AND RELIGION BE INDEPENDENT?

As noted earlier, sometimes people present the idea of strictly limiting science to methodological naturalism by arguing for the *independence* of science and religion. This view intentionally embraces the either/or dichotomy noted earlier, typically saying that science should limit itself to "how" questions that relate to physical causes, while religion should tackle the "why" questions about purpose and meaning. The late paleontologist Stephen J. Gould was a famous proponent of this "non-overlapping magisteria" (NOMA) model in his later years. While this approach sounds elegant and simple, as we have noted it is difficult to put into practice given the significant overlap of science with philosophy and theology when it comes to some "how" and "why" issues. In my own experience, I mostly find scientists espousing the independence model when they want to free science from theological or ethical influence, but ignoring the independence model conveniently when they want science to influence theology and culture. Francisco Ayala illustrates this one-way-street approach well in his book, *Darwin and Intelligent Design*. While giving an overview of the book's contents in his prologue, he comments:

> The final chapter completes my argument by drawing out the important conclusion that evolution (more generally, science) is not incompatible with religion. Indeed, science and religion *cannot* be incompatible, because they concern nonoverlapping domains of knowledge.

[16] See, for example, the blog interview with Michael Ruse, "Is Darwinism a Religion?" *Huff Post* (blog), July 21, 2011, http://www.huffingtonpost.com/michael-ruse/is-darwinism-a-religion_b_904828.html.

But just before that final chapter, when concluding a section that attacks intelligent design theory, he writes:

> The design of organisms is often so dysfunctional, odd, or cruel that it possibly could be attributed to the gods of the ancient Greeks, Romans, and Egyptians, who fought with one another, made blunders, and were clumsy in their endeavors. But for a modern biologist who knows about the world of life, the design of organisms is not compatible with special action by the omniscient and omnipotent God of Judaism, Christianity, and Islam.[17]

Note the striking disconnect here: first we are assured and comforted that "science and religion *cannot* be incompatible, because they concern nonoverlapping domains of knowledge," but later we are told that any modern biologist who knows anything about science cannot believe in a creator god in the traditional sense of the term. Apparently only science and *certain types* of religion are compatible. Moreover, it appears any biologist who disagrees with Ayala's religious conclusions doesn't really understand science.

COLLATERAL DAMAGE

However much methodological naturalism has restricted the physical and biological sciences from being open to the possibility of God acting in and guiding the world, perhaps the greatest damage of this approach has been in the humanities and social sciences. By attempting to be "scientific" and to focus strictly on material cause and effect, people have been dehumanized and reduced to a complex set of chemical reactions. Self-worth, dignity, and responsibility are replaced by genes and environmental factors, so we become mere pawns and victims—unless we imagine ourselves at the top of the evolutionary ladder, in which case we can use our superiority to suppress or eliminate those who are less fit than we are. Although

---

[17] Francisco J. Ayala, *Darwin and Intelligent Design* (Minneapolis, MN: Fortress, 2006), *X*, 89. Emphasis in original.

the focus of this book is on the natural sciences, the application of these naturalistic methodologies in the social sciences has reduced humans to animals. Christians in these disciplines who recognize this error should work to restore the vision that we are, instead, a little below the angels.

WHERE THE CONFLICT REALLY LIES

In summary, the modern tensions between science and Christianity are fundamentally religious ones, but not over what you'd think: the tensions are between Christian and naturalistic worldviews, which generally are separate from the data. Unfortunately this worldview distinction is lost in the cultural prestige of "science," which often presents itself as the neutral pursuit of truth. In origin-related questions, at least, science needs to be free to follow the evidence wherever it leads, instead of restricting itself to tell a naturalistic story no matter what. Despite these worldview differences, there is massive common ground between secular and Christian approaches in the sciences, because both have an interest in physical processes and in finding truth. While the secular approach sees the physical processes as an end in themselves, a Christian in the natural sciences will prefer physical causes for physical effects because God often works through secondary *natural* causes. The Christian enjoys the freedom, however, of not having to force-fit his theories into a naturalistic box, recognizing that the pursuit of truth should not have religious restrictions.

+ 6

BARRICADES

THE END OF THE ROAD FOR REASON AND EXPERIENCE

There are more things in heaven and earth, Horatio,
than are dreamt of in your philosophy.

Hamlet, 1.5.166–67

Let's go back to where our look at the history of science has left us. Fostered in a Christian worldview, the early sciences saw the universe as the handiwork of an ingenious God, who governed his creation through laws that we could discover and understand because we were made in his image. But over time mechanical explanations prevailed, pushing God more and more into the distance until there was no apparent need for him, nor any real sense of his involvement or guidance in the world.

Is this how the story ends? Has science shown us that we are but an accident among the gears of a cold, mechanical, clockwork universe that ticks and grinds along without meaning or significance? Or did we err in thinking that we could remove God from his universe and still be able to make sense of it? There is good reason to think the latter and that the Enlightenment goal of offering a full explanation for the cosmos by referring solely to human reason and experience (known today as methodological naturalism) has failed.

THE LIMITS OF MATHEMATICS

First, there is the problem with human reason itself. Mathematics, seen for millennia as the purest form of logic and the gold standard for truth, turns out to have its limits. One limit was famously proven by Kurt Gödel's incompleteness theorems, which showed that simple logical principles alone could not prove the internal consistency of any mathematical system. (An example of this is the statement, "This sentence is not provable.") In addition to this logical consistency problem, the second dilemma is that there's no reason why mathematics, left to its own devices, so faithfully describes events in the physical world. As famously noted by Eugene Wigner:

> The miracle of the appropriateness of the language of mathematics for the formulation of the laws of physics is a wonderful gift which we neither understand nor deserve. We should be grateful for it and hope that it will remain valid in future research and that it will extend, for better or for worse, to our pleasure even though perhaps also to our bafflement, to wide branches of learning.[1]

This reflection should make us pause. The tool which Galileo saw as the "language of God," which gave Kepler the insight into God's thoughts about the riddle of planetary orbits in the solar system, and which Newton developed in order to unlock God's regular actions, is not as self-evident without God as people thought. To cite Wigner again, "The enormous usefulness of mathematics in the natural sciences is something bordering on the mysterious and . . . there is no rational explanation for it."[2]

If the correspondence of mathematics with reality has no rational explanation, then what can we say about physics and the other sciences? No one doubts that the equations work, but *why* do they work? And why can we figure them out, and why do they

---

[1] Eugene P. Wigner, "The Unreasonable Effectiveness of Mathematics in the Natural Sciences," *Communications on Pure and Applied Mathematics* 13 (1960): 1–14.
[2] Ibid.

make sense to us? Questions like these had clear answers back when science grew within the Christian tradition: God is the Creator. He created the universe to work in regular ways, and we can understand these patterns because we are made in his image. The Enlightenment and positivist goal of removing God from science may have been culturally successful, but it left us with "no rational explanation for it [why mathematics works in the natural sciences]." Perhaps it would be appropriate to acknowledge once again the Giver of this "wonderful gift"? To cite Kepler, "God wanted to have us recognize these laws when He created us in His image, so that we should share in His own thoughts."[3]

THE LIMITS OF NATURAL SELECTION

But the problems with human reason extend beyond the formalisms of mathematics: there is the question of whether the thoughts of a brain that evolved through mutation and natural selection can be trusted. Charles Darwin worried about this himself. In a letter to William Graham on July 3, 1881, he wrote,

> Nevertheless you have expressed my inward conviction, though far more vividly and clearly than I could have done, that the Universe is not the result of chance. But then with me the horrid doubt always arises whether the convictions of man's mind, which has been developed from the mind of the lower animals, are of any value or at all trustworthy. Would anyone trust in the convictions of a monkey's mind, if there are any convictions in such a mind?

Philosopher Alvin Plantinga develops this concern as one of several forceful arguments against naturalism in his recent book, *Where the Conflict Really Lies: Science, Religion, and Naturalism,* where he also presents the case for a strong consonance between theism and science.[4]

---

[3] Quoted in Max Casper, *Kepler*, trans. and ed. C. Doris Hellman (New York: Dover, 1993), 380.
[4] Alvin Plantinga, *Where the Conflict Really Lies: Science, Religion, and Naturalism* (New York: Oxford University Press, 2011).

THE LIMITS OF EXPERIENCE

But it is not only rationality and human reason that fell upon hard times: experience suffered as well. Newton's famous equation of gravity, which allows us to calculate the position of planets and moons with a precision that enables the engineers at JPL to send space probes over vast distances on incredible voyages of discovery, has a problem: it only works exactly when two—and only two—objects attract each other. If you have three or more objects, Newton's simple equations of motion become too difficult to solve because the bodies all attract and influence each other. While this interaction can generally be ignored for short-term solutions, especially for tiny planets orbiting a single large star, such N-body problems can only be solved exactly via computer simulation, where tracking how the forces change as distances vary can be calculated step-by-step over very small time intervals.

While finding a math problem that is too difficult to solve may not be surprising to most readers, it is troubling to physicists, especially when the following fact is added: N-body problems are sensitive to initial conditions. To start a computer simulation, say of the solar system, one needs to input the initial positions of the sun and planets. For a given set of initial values, the equations crank out what appear to be extremely regular and predictable orbits. Determinism is thus proved, is it not?

As it turns out, it's not. If we change those initial values just slightly and run the simulation again, the equations crank out what appear to be virtually identical regular and predictable orbits like before . . . except that after a while they diverge from the first run's results. The subtleties of the planets' interactions with each other eventually cause a significant divergence from the first run's calculated positions, due to those slight differences in the initial values. Thus our tools are inadequate to predict the long-term behavior of a perfectly mathematical, mechanical system. The field of physics and mathematics that wrestles with these problems, known as complexity theory (or chaos theory), mushroomed in the 1970s and 1980s when computers developed the power to run accurate simulations and extrapolate them out to extended time intervals. In some cases, like forecasting the weather, the unpredictability becomes obvious over several days; with planetary orbits it may take millions of years to diverge. But it does. Similar behavior is seen in many physical systems, such as the simple pendulum: in order to solve the equations of motion before the advent of computers, simplifying assumptions were made, and the mathematics gave a solid result. But if you go back and solve for the exact motion without simplifying the equations, you get results that can be chaotic. This suggests that the universe is not as closed and deterministic as we once thought.

THE LIMITS OF KNOWLEDGE

But wait—those chaotic differences appeared when we changed the initial conditions slightly. If we knew the exact initial values for our system, wouldn't this problem go away? It turns out that it is impossible for us to measure the exact values accurately enough. Adding a few more decimal places to the initial values helps preserve the result's similarity for a longer time, but because the differences accumulate exponentially, much higher accuracy does not translate into an appreciably longer time before the solutions diverge.

But why can't we measure those actual initial values accurately enough? All measurements of position and speed, try as we might to make them more and more precise, eventually run up against the Heisenberg uncertainty principle. Because of the fundamental wave nature of particles, Heisenberg showed, we cannot simultaneously know both the position and the momentum (which is proportional to the speed) of any object with unlimited accuracy. Thus there is a fundamental limit in our ability to specify the initial conditions that we need to plug into the equations. In effect, there is a "glass ceiling" above which we cannot predict the behavior of even relatively simple mechanical systems, even when we know their equations of motion.

Thus the Enlightenment and positivist project of explaining the universe without God by appealing only to math, logic, and experiments has found some unexpected barriers: there is no rational reason why math and logic correspond to the physical world, and our lab measurements can never be precise enough to justify determinism or reductionism.

Why isn't this troubling to scientists? Probably because most scientists are pragmatists and don't engage with philosophical issues, even when the problems are relevant to their discipline. After all, you can spend your entire career just focusing on systems and time intervals where the equations work well. Given the hard work and narrow specialization that most cutting-edge research requires, there is little time to reflect on or even care about the broader conceptual problems that crop up within or across disciplines.

My first experience with this pragmatic approach came when I was attending the welcome banquet for incoming graduate students in the physics department at Cornell University. Mixing with faculty over dinner in Andrew Dixon White's former carriage house, one new graduate student (who already had a master's degree from a university in Europe, and who knew a bit more than the rest of us newcomers) sat next to a theoretical solid-state quantum physicist, a very pleasant professor who would later be a member of my doctoral thesis committee. The two got to talking about the various interpretations of quantum physics (the most famous is known as the *Copenhagen interpretation*, but there are around a half-dozen others) and soon got into a loud discussion, which grew into a heated argument as the rest of the hall grew increasingly quiet. Finally, with a bang of his fist on the table, the professor blurted out, "I don't give a #%@&* what it means, I just know that the equations work, and that's good enough for me!" One could have heard a pin drop in the room following this, and it took several minutes for conversations and the atmosphere to return to normal.

As Wigner noted earlier, it *is* an incredible thing that the equations work. But is that really "good enough"? What is an equation's significance and what does it tell us about how and why the universe works as it does? If we may harken back to Newton and the law of gravity, "how" and "why" questions about a law like this require *teleological* or *final cause* answers, and these are something that we can't get a handle on by observing the *efficient cause* or by calculating the phenomena. While scientists may do well at number-crunching, and this certainly is fun and important to do, broader questions like these remain enigmas. It's not uncommon for scientists to brush them aside with, "The universe just *is* this way; now please pass me the calculator." But if the universe "just is," we have no reason to presume that the calculations we do today will work tomorrow, since it is logically possible that the laws of physics could change. To assume that they will stay the same borrows from the Christian intellectual tradition in the sciences: if there is a sovereign Creator behind it and we are made in his image, one may reasonably assume that the universe operates according to fixed laws which are intelligible to us. Such assurance evaporates if our universe is simply the result of a just-right quantum fluctuation. Why should the cosmos be stable and reasonably predictable?

A FINE-TUNED UNIVERSE

But there's more. Not merely have the rationality of mathematics and the preciseness of measurements turned into a quicksand for determinism, but the math and data point to outright spookiness in the universe: they show that our universe hasn't always been here, and that the properties of the cosmos are strangely fine-tuned to permit the existence of complex life.

Traditional materialistic philosophies require that the universe be eternal, because any other answer to the question, "Where did matter come from?" suggests that someone immaterial was around to make the material. When Aristotle was reintroduced in the West about a thousand years ago, this became a point of contention between theologians and the Aristotelian-influenced academy: the Bible is pretty clear that the universe has a beginning, but Aristotle, following common ancient tradition, taught that matter had always been here. Thus began one of the biggest points of tension between philosophers and theologians that lasted until about 1930, when Edwin Hubble presented the first clear astronomical evidence for an expanding universe. The most logical explanation for this expansion was that the universe was all wadded together at some point in the past, that it had a beginning and then flew apart. What became known as the big bang theory met with strong resistance *from scientists* because of its theological implications, until additional astronomical data gathered in the 1960s and 1980s ruled out steady-state models and other proposals. The agnostic astrophysicist Robert Jastrow summarized the impact of these findings well:

> For the scientist who has lived by his faith in the power of reason, the story ends like a bad dream. He has scaled the mountains of ignorance; he is about to conquer the highest peak; as he pulls himself over the final rock, he is greeted by a band of theologians who have been sitting there for centuries.[5]

AVOIDING THE OBVIOUS?

Once the idea of a beginning to the universe could not be avoided, after a grudging nod to theologians for being right after all, materialists proposed a universe-generating machine (instead of a universe-generating person, strangely enough) as the source for our universe—and many others! After all, materialists began to argue, it would be strange if we were a unique creation, so we must be one

[5] Robert Jastrow, *God and the Astronomers*, 2nd ed. (Reader's Library, 1992), 107. Jastrow gives a very readable account of the development of the big bang theory and the resistance to its theological implications.

among what could be an infinite number of other universes. This ad hoc *multiverse* hypothesis resolved another embarrassing issue which was cropping up: as theorists developed models for how our universe formed, they discovered an extreme precision in its initial state, expansion rate, mass density, level of density fluctuations, physical constants, and other critical factors which were absolutely essential in order to form a universe that has the right structure and chemistry to allow for the existence of complex life somewhere within it. Known today as the *anthropic principle*, many people at first dismissed this precision with the shrug, "Well, if it hadn't turned out this way, we wouldn't be here to marvel at it." Such a breezy answer was easy for the first surprising factor or two, but today there are over thirty factors, some of which require fine-tuning to levels of 1 part in 10^{60}, making it difficult to discount these coincidences as a selection effect.[6] Philosopher John Leslie relates the growing significance of these "coincidences" by making the analogy to a prisoner facing a horde of sharpshooters, who somehow all miss their target. Afterwards the prisoner remarks, "If the thousand men of the firing squad hadn't all missed me then I shouldn't be here to discuss the fact, so I've no reason to find it curious."[7] Had the prisoner only missed one bullet, such a cavalier response might be possible, but since he missed one thousand, his answer is absolutely incredulous. Similarly, when we look at the universe around us and wonder why we are here, science shows us that the odds are so incredibly low that coincidence is virtually impossible. The better conclusion is conspiracy: someone wants us to be here.

IS IT ALL A HALLUCINATION?

But the multiverse model provides little comfort as an explanation for the fine-tuning of our universe: philosophers have noted

[6] Hugh Ross, "Anthropic Principle: A Precise Plan for Humanity," *Reasons to Believe,* January 1, 2002, http://www.reasons.org/articles/anthropic-principle-a-precise-plan-for-humanity.

[7] John Leslie, "Anthropic Principle, World Ensemble, Design," *American Philosophical Quarterly* 19, no. 2 (April 1982):141–51.

that it is infinitely more likely that a universe-generating machine would create disembodied "Boltzmann Brains" that might *hallucinate* that they are persons with bodies that exist in a universe such as ours, than that we actually exist in a physical universe such as ours.[8] In other words, it's far more likely that you're not sitting and reading this book, but that you are just a brain in a vat in another universe, *imagining* the world around you where you're reading this book. Oh, and all of your past, your family, and your friends are just hallucinations as well. If the multiverse model is true, it requires a vast, blind leap of faith to think that what you see around you is actually real.

Let's review for a moment: the Enlightenment and rationalist movements began in the late seventeenth century because some people thought that the universe looked too mechanical to have God influence it and because some others didn't want him around. Truth, we were told, could be established by human reason and experience alone, without any appeal to ancient texts or religious authority. But in the last century our reason and experience have discovered some clear fingerprints of God in both the structure of the cosmos and its mathematical openness. And yet people are willing to forgo reason and experience by imagining that we are part of a multiverse where anything goes, and which can only be justified with mathematical models that ultimately have no rational basis for why they work. Why is the notion of a Creator so scary that people are unwilling to reconsider the validity of naturalism?

BIOLOGY: THE COMPLEXITY PROBLEM

The world of biology isn't looking terribly comfortable for naturalism either. Although Michael Polanyi pointed out in the 1960s that life did not appear to be reducible to the laws of physics and chemistry (any more than the text of a book or workings of a

---

[8] William Lane Craig, *Reasonable Faith: Christian Truth and Apologetics*, 3rd ed. (Wheaton, IL: Crossway, 2008), 148–49.

machine could be explained only by these laws) such thinking was ignored early on because we didn't understand very well how cells and organisms work. Our knowledge of the complexity of living systems, however, has exploded over the past decades. First we discovered that cytoplasm, the jelly-like substance inside of cells, contained sophisticated, tiny organelles. Then the structure and function of the DNA molecule was deduced, showing it to be the master "cookbook" for making the proteins in the cell. Next came the study of proteins, with their twisted-spaghetti-like shapes and incredible range of catalytic abilities. Proteins often organize into complex structures, like the ribosome, which functions as a molecular machine to transcribe RNA into protein. In fact, almost every significant process in a cell involves molecular machines; rare is the protein that normally works in isolation.

It turns out that proteins are just the beginning: cells contain sophisticated regulatory systems that control when, where, and how much of a particular protein is made. Moreover, these systems involve parts of the DNA that are not protein recipes; the DNA cookbook also contains the calendar and supplemental notes for when and where to make proteins. Recently the ENCODE project reported that over 80 percent of the DNA in the human genome has biological function,[9] ending the notion that most of the human genome is a junkyard where mutations play around and occasionally produce something new and innovative to fuel the advancement of the species.

A FROZEN ACCIDENT THAWS

Then there's the language that the DNA uses for its protein recipes. Called the "genetic code," it specifies how cells convert triplets of nucleotide bases into specific amino acids. Since there are

[9] The ENCODE Project Consortium, "An integrated encyclopedia of DNA elements in the human genome," *Nature* 489 (2012): 57–74. See also M. Kellis et al., "Defining functional DNA elements in the human genome," *Proceedings of the National Academy of Sciences of the United States of America* 111 (2014): 6131–138.

four different types of nucleotides found in DNA, there are 4 × 4 × 4 = 64 possible triplet codes, which specify the twenty amino acid options used in proteins, along with "start" and "stop" commands. When the genetic code was first discovered, Francis Crick called it a "frozen accident," because once this translation table was selected during the origin of life process, any later substantial change to it would dramatically alter the proteins produced and kill the cell. Since virtually all organisms have the same genetic code, this was considered to be very strong evidence for the common ancestry of all life and for the point that this code could not have evolved or improved over billions of years; this DNA-to-protein language essentially had to be fixed at the time of the hypothesized last common ancestor to all life today, or else it would not be so universal. Recent work has shown, however, that this code is hardly an "accident," since the code appears to be almost optimal at preventing proteins from changing, should a random mutation occur in the DNA. In other words, the genetic code is among the most highly error-tolerant (mutations-resistant) codes possible.[10] How this valuable feature could have arisen "by accident" has yet to be explained, although an error-tolerant code makes perfect sense if there's a Creator behind it. Given the importance of error tolerance for stabilizing a species, it is even reasonable to suggest that the common genetic code better reflects the constraints of a common design plan instead of common ancestry.

MULTIFUNCTIONAL GENES

And then, there are the protein recipes themselves. We would expect that one gene would have one recipe to make one protein, and such is the case in simple bacteria. But in viruses and eukaryotic cells (the kind that make up higher organisms like us) there are many situations where one gene contains the recipes for several proteins! Regulatory

[10] Fazale Rana, *The Cell's Design: How Chemistry Reveals the Creator's Artistry* (Grand Rapids, MI: Baker, 2008), 170–82.

mechanisms in the cell, which presently are poorly understood, select which protein from the gene's repertoire is actually made, depending on factors like the type of tissue the cell is in, or the stage of development of the organism. This "alternative gene splicing" vastly increases the complexity and flexibility of the cell; for example, the twenty thousand genes in the human genome are estimated to be able to make over five hundred thousand different proteins.

CELLS ARE . . . COMPUTERS!

Equally fascinating has been the discovery of *gene regulatory networks* that control the growth and development of an organism from the single-cell egg to the adult form. A complex set of messenger RNAs (mRNAs) and proteins interact to guide cell division and cell specialization, so that cells become specific tissues and organs at the correct body location. The level of control and feedback required is such that Eric Davidson, a leading researcher in this field, remarked, "What emerges is almost astounding: a network of logic interactions programmed into the DNA sequence that amounts essentially to a hardwired biological computational device."[11] Elsewhere he comments,

> Neo-Darwinian evolution is uniformitarian in that it assumes that all process works the same way, so that evolution of enzymes or flower colors can be used as current proxies for study of evolution of the body plan. It erroneously assumes that change in protein coding sequence is the basic cause of change in developmental program; and it erroneously assumes that evolutionary change in body plan morphology occurs by a continuous process. All of these assumptions are basically counterfactual. This cannot be surprising, since the neo-Darwinian synthesis from which these ideas stem was a pre-molecular biology concoction focused on population genetics and adaptation natural history, neither of which have any direct mechanistic import for

[11] Eric H. Davidson, *Genomic Regulatory Systems: In Development and Evolution* (San Diego, CA: Academic Press, 2001), 54.

> the genomic regulatory systems that drive embryonic development of the body plan.[12]
>
> These [body plans] are created by the logic outputs of network subcircuits, and in modern animals these outputs are impervious to continuous adaptive variation unlike genes operating more peripherally in the network.[13]

"Logic outputs"? "Network subcircuits"? "Programmed"? No, I did not change the subject to computer science; we're still talking about biology. But you get the idea: we once thought that life was just chemical reactions, but it turns out that life, at minimum, consists of *information processing* in "hardwired biological computational device[s]." Moreover, the major animal types themselves are "impervious to continuous adaptive variation," although small peripheral changes are possible.

NATURAL "MAGIC" OR GOD'S DESIGN?

But Darwin had argued that all life could develop *without guidance* from a common ancestor by numerous, successive, slight modifications. While the mechanisms of mutation and natural selection can generate minor adaptive changes (microevolution), Davidson and others are finding that macroevolutionary changes (in body plan, organs, etc.) require the reprogramming of complex biological logic circuits—a level of sophistication which no one anticipated. It is thus reasonable to ask whether the best explanation for these biological hardwired information processing systems is a Creator, or if we should only marvel over it as "natural magic," as theoretical biologist Stuart Kauffman does.[14] I think it is more reasonable to pause, like King David, and exclaim:

---

[12] Eric H. Davidson, "Evolutionary Bioscience as Regulatory Systems Biology," *Developmental Biology* 357, no. 1 (September 2011): 35–40, http://dx.doi.org/10.1016/j.ydbio.2011.02.004.
[13] Ibid.
[14] Stuart Kauffman, "The End of a Physics Worldview: Heraclitus and the Watershed of Life," *Cosmos & Culture* (blogs), *NPR*, August 8, 2011, http://www.npr.org/blogs/13.7/2011/08/08/139006531/the-end-of-a-physics-worldview-heraclitus-and-the-watershed-of-life.

For you formed my inward parts;
you knitted me together in my mother's womb.
I praise you, for I am fearfully and wonderfully made.
(Ps. 139:13–14)

THE FIRST LIFE?

While biology mainly focuses on living things, one branch of biochemistry considers the question of where the first life came from. The challenge here is to create an initial set of chemicals that can reproduce themselves, since without replication, the Darwinian process of mutation and natural selection cannot work to improve them. But the problem appears to be insurmountable: the minimum chemical complexity for the simplest "life" seems to require about 250 primitive enzymes, which is far too sophisticated to have arisen from lightning bolts striking a warm little pond on the early earth. While hypotheses like an early "RNA world" are often mentioned in the media, the fragile nature of these chemicals and the level of information that they needed to contain in order to replicate themselves makes the origin of life itself almost look like a miracle.[15]

THE LIMITS OF SCIENCE

Charles Darwin, T. H. Huxley, and other nineteenth-century scientists taught us that in order to be "scientific," we must exclude God from our explanations. But doing so over the past century has left us with no basis for why mathematics explains the physical world and left us with the most likely "physical world" being merely the

[15] "An honest man, armed with all the knowledge available to us now, could only state that in some sense, the origin of life appears at the moment to be almost a miracle, so many are the conditions which would have had to have been satisfied to get it going. But this should not be taken to imply that there are good reasons to believe that it could not have started on the earth by a perfectly reasonable sequence of fairly ordinary chemical reactions. The plain fact is that the time available was too long, the many microenvironments on the earth's surface too diverse, the various chemical possibilities too numerous and our own knowledge and imagination too feeble to allow us to be able to unravel exactly how it might or might not have happened such a long time ago, especially as we have no experimental evidence from that era to check our ideas against." Francis Crick, *Life Itself: Its Origin and Nature* (New York: Simon & Schuster, 1981), 88. Note that Crick is assuming a "naturalism of the gaps" approach here.

hallucination of a brain in a vat. Moreover, we found that the universe appears to be open to more than absolute mechanical determinism, and there is strong evidence that our universe is crafted in an extremely unlikely, incredibly precise way so that we could be here. And modern biology shows us levels of integrated complexity that makes the rest of the universe look easy. To explain away these molecular machines and information-processing networks as the result of the unguided and undirected processes of mutation and natural selection seems as incredulous as offering someone a snow shovel to clear off Antarctica. Over the past century we've discovered that the universe and life itself are far more sophisticated than methodological naturalism has the power to explain. Yet the best "scientific" answer is that somehow we're lucky enough to be the one universe where a pile of sand got hit by enough lightning bolts in just the right way to turn into a smartphone. Has the scientific quest for truth left us with "natural magic" as the best explanation? Or perhaps there's something wrong with our definition of "science?" Could the data from the natural world be pointing to its Creator?

Reason and experience are indeed powerful tools to focus our attention on mechanical cause and effect. But when we limited the direction they could lead us, to purely physical causes, it appears that they have brought us to a barricade on the path of knowledge. Perhaps it is time to reevaluate this restricting definition of science and return to the richer set of assumptions available through the Christian intellectual tradition in the sciences.

 7

THE WAY HOME

THINKING OUTSIDE THE BOX OF NATURALISM

A little philosophy inclineth man's mind to atheism, but depth in philosophy bringeth men's minds about to religion.

Francis Bacon, *Of Atheism*

If everything that we see around us is most likely to be a hallucination, then it is difficult to justify doing science, or anything else. It doesn't help when a Nobel Prize-winning physicist comments that, "The more the universe seems comprehensible, the more it also seems pointless,"[1] and that "The effort to understand the universe is one of the very few things that lifts human life a little above the level of farce, and gives it some of the grace of tragedy."[2] These hardly seem like words that will motivate future generations of scientists. Sure, science can give us some fame and fortune, faster computers, and a military advantage, but I think that if Christians, at least, can recover their intellectual tradition in the sciences, it would truly reinvigorate the study of nature and lift the human spirit. The following points are my suggestions for what this involves.

[1] Steven Weinberg, *The First Three Minutes: A Modern View of the Origin of the Universe*, 2nd ed. (New York: Basic, 1993), 154.
[2] Ibid., 155.

1. EVERYONE SHOULD RECOGNIZE THAT EVERY SCIENTIFIC THEORY INCLUDES A PHILOSOPHICAL PERSPECTIVE

One day a student came to my office, saw my diploma on the wall, and exclaimed, "Wow, I didn't know that you had a doctorate in philosophy!" I explained that the "PhD" title goes back to the days when the sciences were called "natural philosophy," thus a "Doctor of Philosophy" is the classic name for an advanced degree in physics and many other scientific disciplines. But, despite the "philosophy" title, scientists today are not trained in philosophy, which creates a problem because they often are unaware of the assumptions that influence their work. As we saw, methodological naturalism involves the prior commitment to explain everything in naturalistic terms. Given that priority, we have good reason to question whether every "scientific" answer is actually true, especially in the areas of origins, history, and theology. Perhaps "science" is simply the best naturalistic just-so story to explain away the obvious design. As we noted earlier, even some atheist philosophers like Thomas Nagel are bothered by the bias which naturalistic religious assumptions have on science and on our culture.

A classic adage about computers is, "Garbage in, garbage out." A great tool, if given bad data, will inevitably give bad results. The problem with contemporary science is similar, except that it is more like a warped tool. Any data it is fed, no matter how suggestive of design it is, will generate a naturalistic answer. Given its prior philosophical commitment to naturalism, statements like "Science cannot detect God" are not surprising. Yet the fine-tuning of the universe seems to suggest that we are seeing the results of God's choices and guidance. And the complexity of life reflects precisely the engineering prowess that minds are capable of and dice-throwing is not. Because the Christian intellectual tradition seeks truth with no holds barred, it is open to a broader range of causal explanations than naturalism.

While openness to more than material explanations is the biggest factor, there are more subtle issues where philosophy also comes into play, such as the fact/value dichotomy. Science, we are told, gives us "facts" about the world, whereas religion gives us "values." Of course, this means that science deals with objective information and religion with subjective feelings and experiences. From here it is a simple step to dismiss Christianity entirely when it speaks to historical and origin questions, leaving science as our only guide to truth in these areas. Science today, however, is filtering its data through a naturalistic filter when it produces "facts," and the Bible is rarely subjective: it includes quite a few empirical claims about the human condition, as well as statements about origins and history (see the apostle John's empirical claims about the risen Jesus in 1 John, for example). If scientists were more aware of the history and philosophy of science, they would see that the fact/value dichotomy is too simplistic.

The origin of life and macroevolution are two areas where philosophical assumptions dramatically influence scientific theory. Origin-of-life research generally concentrates on figuring out how the raw ingredients came about on the early earth, but this material focus on amino acids, nucleic acids, sugars, and lipids misses the information problem. Raw ingredients are necessary, but not sufficient, to bake a cake. Attempting to synthesize life from the preexisting parts of other organisms is certainly an interesting and challenging biochemical research project, but it tells us little about life's actual origin, which is fundamentally an origin-of-information problem. Over the past decades, we have come to realize that life is information processing and much more than, not mere, chemistry. Yet an absolute commitment to a "naturalism of the gaps,"[3] instead

[3] My use of this phrase here is to point out that everyone bridges their gaps in knowledge with their worldview assumptions. The derogatory phrase "God of the gaps" is a common accusation against creationists, but those who consider God to be part of the explanation are not arguing that the matter cannot and should not be studied further. To achieve his ends, God often uses means that are open to our investigation and application.

of following inference to the best explanation, stifles research in this area.

Strong theological assumptions are equally obvious with macroevolution, and the history of these goes back to Darwin. A very common argument for evolution from Darwin's time forward is that "God wouldn't do it that way"; in other words, the animals we see must have evolved because if God had created them, he would have made them more efficiently, benevolently, or wisely. Try as we might to protect God from getting the blame for apparent bad designs, this either-God-or-evolution argument doesn't work: first, if God exists, then he designed the evolutionary mechanism, so he is still responsible for its outcome; second, God actually accepts responsibility for the bad designs (recall our discussion of Exodus 4:11), so we do not need to defend the Victorian image of an omnibenevolent God who created the world for our happiness. Apparently the things that we think are bad designs or mistakes are serving an important purpose in this world, which was created to be the final battleground of good versus evil.

2. WE NEED TO PRACTICE GREATER CHARITY AMONG THOSE WHO HOLD A VARIETY OF VIEWS

As a Baptist who attended a Presbyterian seminary, I am used to long discussions over the most biblical way to baptize someone, whether we should baptize only confessing adults, what is the most biblical means of church governance and organization, and other denomination-related topics. These sorts of differences of opinion have, over the history of the church, given rise to the range of affiliations that we see today among Protestants.

So it's not surprising that there will be differences of opinion among Christians about how to deal with interpretative issues regarding the sciences and the Bible; however, I am disappointed that they are often so militant and ad hominem. I've met theistic evolutionists who despise any type of creationist, and young-earth creationists who believe that anyone who doesn't agree with them isn't really a Christian. I suggest that some reconciliation before we get to heaven might be appropriate.

Before continuing, it will be helpful to map out the spectrum of Christian views regarding origins:[4]

A. *Young-Earth Creationist*. This view holds that the earth and universe were created in six twenty-four-hour days a few thousand years ago, with a special creation of Adam and Eve.
B. *Old-Earth Creationist*. This position argues for a six-*era* creation of the earth and universe over billions of years, with a special creation of Adam and Eve around 50–100 thousand years ago.
C. *Theistic Evolution*. This term applies to a range of views which hold to an old earth and universe but do not hold to the special creation of Adam from nonliving material and Eve from Adam's side. Roughly these are:

1. *Directed Evolution*. Does not read Genesis 1 as a sequence of historical events and thinks that God guided physical and evolutionary processes without distinct interventions. Adam and Eve evolved from early hominids, but were a historical couple ensouled by God to have a special relationship with him.
2. *Planned Evolution (Evolutionary Creation)*.[5] God created and set up the universe so perfectly that his later intervention and guidance were not necessary in order to accomplish his goals. Humans evolved, but not from an original pair. Genesis gives us no scientific insights.
3. *Nonteleological Evolution*. God created the universe and has not intervened or guided it since, allowing it to run via undirected natural processes. Humans evolved, and Genesis is a Hebrew myth about origins. This view differs from planned

[4] For a good review of these positions, see Gerald Rau, *Mapping the Origins Debate: Six Models of the Beginning of Everything* (Westmont, IL: InterVarsity, 2012). I have summarized his categories here.

[5] Some prefer the term "evolutionary creation" to "theistic evolution," noting that they want to place the emphasis on creation instead of evolution. However, among theologians the term "creation" is reserved for views which hold that God created Adam from nonliving material (taking "dust of the ground" in a literal sense). Since "evolutionary creationists" believe that Adam evolved from hominids, their use of the "creation" noun is thus confusing.

> evolution chiefly over whether God has any specific plan for the universe.

As you can see, there is considerable range here among Christian positions, from a total acceptance of naturalistic/scientific explanations (with skepticism about the relevance of the Bible on origin questions) to a total acceptance of the Bible (with skepticism about the relevance of naturalistic/scientific explanations regarding origin issues). While everyone recognizes microevolution, limited adaptation, and that the accumulation of mutations leads to the degradation of a species over time, theists differ on how much complexity and new information can be produced without God's guidance and/or intervention in the process. You also see that intelligent design (ID) does not appear as a distinct category, because it is not. Despite its being blasted in the media as a fundamentalist creationist position, ID only makes the claim that God's design in nature is detectable by science.[6] Thus it is compatible with, and has proponents who hold to, the directed and planned evolutionary positions as well as the young- and old-earth creation views.

While there are minor variations within the young-earth creationist position and the old-earth creationist position, the theistic-evolutionist group is the most diverse. The common ground among this group is that they believe that Adam and Eve were not special creations physically, but that they evolved from early hominids. Beyond this point, whether evolution is a guided or unguided process, whether the soul is a spiritual quality distinct from the physical body, and whether early Genesis offers any scientific or historical insights are difficult to generalize.

To help sort this wide range of positions out, I've found help from philosophers, who recognize that every discipline faces both internal and external conceptual problems. Theology focuses on

[6] A good introductory guide to this position is William A. Dembski and Sean McDowell, *Understanding Intelligent Design: Everything You Need to Know in Plain Language* (Eugene, OR: Harvest House, 2008).

the Bible and tries to come up with an internally consistent way to understand the text and God. The sciences study the physical world and try to come up with an internally consistent explanation for how the universe works. There are topics, however, where both disciplines speak to the same questions, such as how the universe came to be or how humans got here, that can give rise to external (or interdisciplinary) conceptual problems if they offer different answers. In such cases it's very easy for a person who is well-versed in the Bible to be suspicious of science, since he's not that familiar with it and has reservations about its naturalistic leanings; likewise, it's very easy for a scientist to wonder if the Bible is at all relevant to origin questions since it is an ancient text. Christians will tend to gravitate toward conclusions in the spectrum that best fit within their area of expertise.

While these trends are understandable, I encourage readers on all sides to think outside their comfort zone, and certainly to be friendlier when talking about this subject. From looking at the past, we see that there was variety in the Christian intellectual positions in the sciences, reflecting leanings that were Aristotelian, Neoplatonic, and/or mechanistic. Yet each perspective brought a fresh insight to our study of nature, and in retrospect, each turned out to be partially correct. Aristotle's organic view works well in biology, and Neoplatonists anticipated the inherent forces of gravity and electromagnetism along with their mathematical elegance, while mechanical philosophers were correct that our bodies are composed of atoms.

I have always found interdisciplinary discussions extremely fruitful, alerting me to question and justify the things I take for granted or directing me toward solutions I would not otherwise have considered. But wading through the "those guys are idiots" and "those guys aren't really believers" attitudes and speeches is tedious. Theistic evolutionists who think that creationists are idiots should realize that their secular colleagues think that *all* Christians are idiots because we believe that dead people come back to life. And come to think of it, if God had the power to raise Jesus from the dead, is it really that much of a leap to think that there was a historical Adam? Or even that God would create Adam from scratch? If we worship a God who blew out galaxies like dust in astronomy, why do we stifle his creativity and abilities in biology? On the other hand, for my young-earth creationist friends who doubt whether those who disagree with them are really believers, I suggest that perhaps they have made the gate too narrow. After all, we are *interpreting* the Bible, and there is no answer key in the back of the book to assure us that we have gotten precisely the right answer. Do we really need to dismiss so much of the science that suggests the earth is very old, in order to maintain this particular interpretation?

For the sake of full disclosure, I should mention that I personally think that an old-earth creationist perspective, as reflected by groups like Reasons to Believe (www.reasons.org) and the Interdisciplinary Biblical Research Institute (www.ibri.org), offers a good approach for reading the biblical creation accounts as historically accurate, while respecting the scientific data we have gained about the cosmos. I find that this model has traction with nonbelievers while avoiding the "either science or the Bible" dichotomy that tends to happen among theistic evolutionists or young-earth creationists. Of course, I must ask my own position hard questions like: Am I reading too much science out of (or into) the text of Scripture? Are the details about the creation of Adam and Eve and their history critical points of Christian belief or not?[7] If so, how do I harmonize that theological view with the scientific consensus about human origins? Is the consensus *that* blinded by naturalism?

Historically, we have seen that Christians took a variety of approaches to imagining the world and God's relationship to it.

---

[7] For a good treatment of this question, I would recommend C. John Collins, *Did Adam and Eve Exist? Who They Are and Why You Should Care* (Wheaton, IL: Crossway, 2011).

Many times these differences led to heated arguments. It would be simplistic to presume that unity is possible today or that clashes will fade. More importantly, we also saw that each of these approaches had strengths and value at providing fresh insights and in asking important questions—and it turned out that each early view had a piece of the answer. Thus, as the Christian intellectual tradition in the sciences moves forward, it is worth valuing a range of positions, recognizing that two centuries from now, if the Lord tarries, scholars will likely note that each view had part of the answer for today's controversies.

While recognizing the richness of a range of Christian views on origins, I would like to add one personal cautionary note: I am concerned that the planned and especially the nonteleological evolutionary views may be too far afield. Given the powerful fine-tuning arguments coming out of astrophysics today, it seems unnecessary to continue to insist that God never did anything that science can detect. And if we see God's hand in astrophysics, but then deny that God could have guided biology, it seems to imply that God was a great physicist but he didn't take any interest in living things. For me, at least, that would be strange, considering that he became one (the incarnation) and considering that living cells are more complex than the rest of the universe. Also, many who take these naturalistically heavy positions say that we need to accept "consensus science" and change our theology to match it, so that we will be more appealing to non-Christians. I fear that this approach misses the fact that there are very strong theological currents in the sciences, especially over questions about origins, and many "scientific" conclusions on origins ultimately hinge upon what people think God would or would not do, not upon the data themselves. While I certainly think that we need to present as credible a "mere Christianity" position in witnessing to a secular world as we can, it is possible to concede too much and end up with virtually a deistic worldview instead of a theistic one. To me, mere

Christianity should include an acknowledgment of God as Creator as well as humanity's rebellion against him, and mention that the precise details of these events are obscured by their distance in the past so there are a range of possible interpretations about their history. I also think it is an important part of the Christian tradition to convey God as active in the world today, caring for his creation and answering prayer. Such an approach—acknowledging genuine differences but maintaining core beliefs—would provide us a common voice to reach out to the lost without fighting with each other. Of course, in-house discussions over the blinding effects of naturalistic assumptions and how literally to interpret Scripture should continue, like my Baptist-Presbyterian ones did, but with warmth instead of heat.

For non-Christians who might be reading this book, I invite you to recognize that greater charity from those on your side regarding non-materialistic positions would be helpful too. I've read too many secular critics who write with a religious furor in apocalyptic tones about how science and even civilization itself will collapse if creationists are shown any respect whatsoever. Frankly, this often feels akin to racism. Any fair look at the history of science shows that Christians and theists have been major players and contributors to the field (unless one doesn't consider James Clerk Maxwell, Michael Faraday, Isaac Newton, Johannes Kepler and others to be "scientists"), and there is no reason to expect this not to continue.

I recall being asked one afternoon by my advisor in the lab, "So what would you do if you discovered something here which disagreed with your religious views?" His tone was friendly, and thankfully, I'd thought about this question already, so I replied, "Well, I worship a God of truth, so I'm not expecting to encounter any problems, since we're seeking truth here in the lab. But if I did encounter a difference, I'd have to weigh my lab results with my theology, to see how they could better fit." My advisor was happy with that answer, since he'd told the group many times that his goal was truth, not conformity to any accepted models. In fact, at a reception many years later, he mentioned that the high points of his diverse and esteemed research career were the times when he demonstrated that an "accepted model" was wrong.

People who bring different assumptions and perspectives to the lab can raise interesting research questions. For example, it is universal in biology today to assume the common ancestry of all life, so that all genetic, biochemical, and anatomical data are automatically fit into that model. Data that don't fit this model, however, are considered a "research problem" or "noise" and are rarely discussed, outside of some speculation about "convergent evolution." But if common ancestry was not the only game in town, what would the data actually show? Could the data dismissed as "research problems" yield insights that lead to fresh discoveries?

Generally an evolutionist asks, "What is nature capable of doing?" and takes a bottom-up approach to the riddle of life, seeking blind mechanisms to go from molecules to man. Design is therefore unexpected and illusionary. A creationist will ask, "What did God do?" and take a top-down approach, looking for purpose and design in organisms. Figuring out the precise mechanisms that God used are not too important to a creationist because a designer may work within the context of natural laws (providence), or he could input information or structure directly. Since we don't know the mechanisms by which humans create new music or a computer code, it shouldn't be surprising that we can't specify exactly the mechanism by which God created.

But note that the complex organization and the information processing in living systems, and the fact that we need to reverse engineer them to understand how they work, are all *expected* in a creationist approach to science. These are *surprising* to a naturalist, whose default expectation is simple tinkering and chaos. Thus a variety of approaches brings a broader range of questions and expectations into the lab, and may spur research in areas that might otherwise be ignored. For example, it was creationists who were among the first to challenge the idea that "junk DNA" was nonfunctional, a concept that stifled genetic research for decades.

Besides weighing the benefits that multiple approaches can bring to solving problems, I also invite naturalists to consider the question, "How would you know if purely naturalistic mechanisms were inadequate to explain the universe?" When I've asked this question of secular scientists in friendly personal conversations, I have gotten stunned looks: this idea has never been thought of, let alone thought through. Of course, looking for natural causes for physical effects is a good place to start and has been an important tool for the advancement of scientific knowledge. But as important as a hammer is, you can't build a house using only a hammer—you need a suite of tools to construct it well. Perhaps more than natural causes were necessary to make our universe? If so, wouldn't it be "scientific" to know that? What signals in the data would point us in that direction?

Although methodological naturalism is an important component of the scientific method, we all need to be aware of its metaphysical implications and avoid concluding that the universe is purposeless because science says so. To make that assumption in order to simplify our study of nature is one thing; to turn that assumption into a conclusion is a logical fallacy. I suggest the best way to avoid such errors is to allow more worldview diversity, especially in biology. As Newton noted, there is a vast ocean of truth yet to be explored, while we busy ourselves playing with some pebbles here and there on the seashore. A range of worldviews and perspectives brings insights and questions to the table which might not otherwise be considered. The greatest breakthroughs in science generally come when people break free from the herd and think outside the box. Friendly and thoughtful interaction will help science progress.

3. WHEN DOING SCIENCE, AS IN THE REST OF LIFE, CHRISTIANS NEED TO KEEP GOD IN THE PICTURE

One of the biggest challenges that the early church faced was the heresy of gnosticism. This view held that the material world was evil, created through error and ignorance, and that only the spiritual world was good. Thus the things of this world were despised, while mystical and religious experiences were sought to give a glimmer of the heavens to which the soul would eventually return. Against this mind-set the church fathers argued that the world and the flesh were not inherently evil because Jesus was God incarnate. Therefore, the Most High God, not some lower demiurge, really did create the heavens and the earth, and they were good.

It's fascinating that Christians today face almost exactly the opposite problem: a functional materialism, where we focus on the physical world with only the vaguest notion of something beyond it. Some imagine God as a form of energy or some physical substance, or some live their "Christian" lives focused entirely on the material world and its goals, where salvation is some abstract, future fire-insurance policy. Keeping a healthy balance between the physical and spiritual worlds is evidently hard to do.

The biblical picture of God and creation emphasizes two qualities: first, God's *transcendence* over the universe, meaning that he is above, beyond, and separate from the physical world, which he created.[8] Second, God's *immanence* means that he is present in, but not part of, the physical world, and he relates to it intimately. Paul stresses these two attributes of God in his speech on Mars Hill,

> The God who made the world and everything in it, being Lord of heaven and earth, does not live in temples made by man, nor is he served by human hands, as though he needed anything, since he himself gives to all mankind life and breath and everything.

---

[8] For more details, see William Dembski, "Transcendence," *DesignInference* (blog), accessed July 8, 2013, http://www.designinference.com/documents/2003.10.Transcendence_NDOCApol.pdf.

> And he made from one man every nation of mankind to live on all the face of the earth, having determined allotted periods and the boundaries of their dwelling place, that they should seek God, and perhaps feel their way toward him and find him. Yet he is actually not far from each one of us, for "In him we live and move and have our being"; as even some of your own poets have said, "For we are indeed his offspring." (Acts 17:24–28)

Thus the biblical picture of God differs from pantheism, which sees God in everything, and everything as God. In Christian theology God is not part of the creation, yet he is deeply involved with it. As Jesus noted in the Sermon on the Mount,

> Are not five sparrows sold for two pennies? And not one of them is forgotten before God. Why, even the hairs of your head are all numbered. Fear not; you are of more value than many sparrows. (Luke 12:6–7)

It is simply unimaginable to us that God has such a detailed level of knowledge and control of the universe, such that "he determines the number of the stars; he gives to all of them their names," (Ps. 147:4). And yet, as astrophysicists have uncovered the fine-tuning of the universe and biologists have uncovered the programming and engineering inside cells, this biblical picture begins to make sense. Any being who has the power and skill to create a universe with the precision and perfection that is required so that creatures like us can exist in it, would logically have the knowledge and ability to take a personal interest in us individually. If he controls the total amount of matter in the universe to the point that one dime's mass more or less would make complex life impossible,[9] then the universe itself is a physical demonstration of both his transcendence and immanence.

There is a personal message we can take from this "unimagi-

[9]Hugh Ross, "A Dime's Worth of Difference," *Reasons to Believe*, October 1, 2006, http://www.reasons.org/articles/a-dime-s-worth-of-difference.

nableness" as well: God cares for us, and nothing happens to us outside of his will. Of course, God's care for us does not make him a genie in a bottle, or a slave to do *our* will: God's purposes for us will likely include great trials and hardships that we won't understand. But it is incredible to realize that we are not "lost in the cosmos" but are here for a purpose that will ultimately bring glory to him and serve others.

While this big picture is awesome, a question arises when we, as human theologians, philosophers, and scientists, look at the details: How should we think of God's interaction with nature? Historically we saw that Greek influence gave rise to differing approaches: Aristotle thought of the physical world in organic terms, where objects "had a mind of their own," which directed their behavior. Newton and other Neoplatonists thought of the world as completely passive particles, which moved and interacted under the direct hand of God. Our best picture today is that particles exert some inherent forces such as electricity and gravity. So where and how exactly is God involved?

On this question, theologians have tended to make a distinction between *natural* and *supernatural* causes. Natural events occur through the action of regular, law-like processes, while supernatural events involve the direct action of God "above nature," as we see in miracles. Unfortunately, this distinction tends to set up an either/or way of thinking, as we saw earlier, where believers are primed to see God at work only when miracles occur.

The Reformers developed a less misleading way to think about God's actions, distinguishing between *miracles* and *providence*. Miracles are the direct intervention of God, whereas providence involves his guidance and direction of natural events to achieve his intended results. Philosophers describe this approach in terms of God's *primary* and *secondary causation*. Just as I am able to drive an automobile around town without breaking any of the laws of physics, God is able to accomplish many of his purposes through secondary causation, without breaking any of the laws of physics. He uses miracles when he especially wants to get our attention.[10]

Thus it seems that we err as Christians if we leave God at the door of the lab. The Bible paints a holistic picture of God and nature, seeing everything as ultimately under his care. His providential guidance means that we can expect to see regularity, not chaos, around us. Bringing God into the lab leads us to expect to find ingenious patterns and order as we study his handiwork. Those who think that allowing a "Divine Foot in the door"[11] would bring chaos to science have misread the history of science and its Christian intellectual tradition.

In hindsight, it is unfortunate that the early mechanical philosophers used the gears in a clock as a metaphor to describe the regularity of God's creation. Back in the seventeenth century, accurate pendulum-driven clocks were the latest new technology, and were becoming commonplace. (A comparable modern metaphor would be saying, "The universe is like a smartphone.") Unfortunately, skeptics soon capitalized on the weaknesses of this metaphor. Since the best clocks don't require any adjustment from the outside, the clock and clockmaker metaphor was soon adapted to fit deism rather than theism.

Perhaps a better metaphor than thinking of the universe as a clock is to imagine it as a musical instrument: the instrument's structure and behavior follow the laws of physics, but it is designed for an artist to play, not to wind up so that it will endlessly repeat some melody. While the musician does not break any of the laws of physics when he plays it, its sound reflects his personal creativity and skill. Yet without the artist's input, the instrument does

---

[10] For a more detailed discussion, see Thomas F. Torrance, *Divine and Contingent Order* (Edinburgh: T & T Clark, 1998). See also Christopher Kaiser's historically focused discussion of "relative autonomy" in *Creational Theology and the History of Physical Science: The Creationist from Basil to Bohr* (Leiden: Brill, 1997), 32ff.

[11] Richard Lewontin, "Billions and Billions of Demons," review of *The Demon-Haunted World: Science as a Candle in the Dark*, by Carl Sagan, *The New York Review of Books*, January 9, 1997, http://www.nybooks.com/articles/archives/1997/jan/09/billions-and-billions-of-demons/?insrc=toc.

nothing at all interesting. This metaphor incorporates the transcendence and immanence of the musician relative to the instrument itself, while recognizing the regular, mechanical, law-like behavior of the instrument. Perhaps if Robert Boyle had used this imagery instead of that of the clock, it would have been less malleable to a deistic reinterpretation.

As scientists, then, we have the privilege of studying the physical world as God's instrument, noting its regular features and behavior, as well as its ingenious construction. Alongside theologians, perhaps we can strive to hear and understand the melody that God is playing.

4. CHRISTIANS IN THE SCIENCES, LIKE THOSE IN OTHER DISCIPLINES, SHOULD MAINTAIN A HEALTHY REGARD FOR THE SCRIPTURES

As Christians, how should we bring the Bible to bear in our discipline? I speak here more broadly than the natural sciences alone, because this question also impacts historical studies and all other areas, really. To answer in one word: *Appropriately*.

The challenge we face when interpreting the Bible should not surprise a scientist: interpreting the words on a page can be as difficult as interpreting the light from a telescope. Unpacking the information available in the spectra of stars takes hard work and has been improving for a century; unraveling the nuances of a text isn't easy either, although it has more centuries of work behind it. When I walked from a physics lab into a seminary classroom, I found that I needed to bring the same tool kit with me: the skill of weighing the various options and using "inference to the best explanation" to arrive at a conclusion that seemed most appropriate. As our knowledge base changes (I almost wrote "improves" here, but we both gain and lose knowledge over time), our sense for what is the best understanding of a biblical passage may shift. This is the human limitation to understanding nature and Scripture: both data sets are filtered through interpretive lenses to give us science and theology. Given these interpretive layers, how do we integrate them appropriately?

This is not a question which first arose in modern times, and it has received considerable thought from theologians over the centuries. Augustine faced it when integrating Greek philosophy with the Bible around AD 400, and Thomas Aquinas wrestled with the question as he integrated Aristotle's rediscovered works with theology around AD 1200. Both came to the same conclusions about interpreting a biblical text involving the physical world in the context of the then-accepted cultural understanding about the world. These were that (1) the truth of Scripture is not in question, and (2) when several reasonable options are available, no particular interpretation should be held dogmatically as the definitive sense of the text, because a good argument in the future might show it to be false. Both Augustine and Aquinas wanted to avoid the situation where people were asked to believe something that their contemporary science held to be false. Of course, they used this approach sparingly, because they did not want to dismiss the resurrection, for example, because people "knew" that dead bodies didn't come back to life; their main application of this principle was in Genesis and other biblical descriptions of the physical world. Today theologians refer to this interpretive approach as *accommodation*, whereby God spoke through the biblical writers using layman's language in terms of the way things appear, and thus we can err if we take these ancient idioms and phrases too literally (much as we would misunderstand the English words "sunrise" and "sunset" if we took them literally). Thus to integrate appropriately, we need to ask what the biblical text is intending to convey.

We should also avoid a weak view of inspiration, which can surface from attempts to dismiss early Genesis as a sanctified ancient Mesopotamian and Egyptian myth and therefore as irrelevant to any scientific questions about our origins. When I read Christian authors prior to the 1700–1800s, I am often struck by the authority which they gave Scripture; since the Enlightenment, we've been affected by secular pessimism and tend to backpedal when perhaps we shouldn't. Consider, there certainly are interpretive problems in recovering the precise meanings of ancient words, but shouldn't we at least think that early Genesis texts attempt to convey history in simple terms that readers are capable of understanding universally across global cultures over the centuries? Can't God speak truth that is intelligible to his immediate audience as well as to later readers? If pastors can give "children's sermons" that touch the five-year-olds up front as well as the adults in the pews, perhaps God can, too. To integrate appropriately, we need to consider the ability of the Author.[12]

Since Christians take a range of positions on scientific theories like Darwinism, we should not be surprised if Christians take a range of positions on what is "appropriate" integration. But, I suggest that the simplest approach (which is all too common) of dismissing the Scriptures as not having anything relevant to contribute to our modern scientific knowledge is not the correct answer. We should not be surprised to find insights there (such as the concept of a beginning to the universe) if the Author is the God we say we believe he is. While I will leave the detailed coverage of integration in the other disciplines to the other books in this series, the Christian should think about how God influences history, and about biblical insights into human nature, philosophy, music, art, and . . . everything. To presume we are "beyond the Bible" would be casting ourselves adrift not only from a great intellectual tradition, but from our Creator's message.

---

[12] Theologian C. John "Jack" Collins does an excellent job working through the basic interpretive factors for understanding early Genesis in his *Genesis 1–4: A Linguistic, Literary, and Theological Commentary* (Phillipsburg, NJ: P & R, 2006). His *Science & Faith: Friends or Foes?* (Wheaton, IL: Crossway, 2003) is a more layman-accessible review of these questions, offering more interaction with science. Physicist Hugh Ross presents a detailed integrative model in *Navigating Genesis: A Scientist's Journey through Genesis 1–11* (Covina, CA: Reasons to Believe, 2014).

5. CHRISTIANS IN THE SCIENCES, AS IN THE REST OF LIFE, NEED TO DEMONSTRATE CHRISTIAN CHARACTER

While it is rare at Biola University, I am always deeply troubled when I find a student cheating on an exam or term paper. Christians are supposed to be above that: Jesus didn't cut corners, and he said that he was "the way, and the truth, and the life" (John 14:6). As his followers, we should strive to be honest.

Out in the secular market and workplace, people know this, too, and hold Christians to a higher standard. I've seen this happen both ways, as a negative rebuke, "I thought you were a Christian, but you're no different from anyone else!" and in positive shock, "Why are you so honest?"

In the sciences, truth and honesty are critical. Teamwork, collaboration, and the ability to base new research on previous work, all demand that scientists be accurate and honest. If scientists cannot trust each other's results, then the "many hands make light work" approach fails and progress is hindered. As Christians who worship a God of truth, we need to pursue truth, even when that comes at a cost (not getting as good a grade as others who cheated or not getting funding because we didn't "fudge" our results to appear successful).

Christians have been negatively stereotyped by today's popular media and we need to work extra hard to break those stereotypes. One of my students encountered this when she took a summer research position at a major university in the Los Angeles area. At her first group meeting, my student introduced herself by saying that she attended "the Bible Institute of Los Angeles." She was surprised when the room "got quiet." I suspect that nobody in the room had ever met a real Christian before, and their reaction came from negative impressions they had picked up from sitcoms and other media that convey us as bigoted, judgmental, narrow-minded idiots. The solution to this is not to apologize for our beliefs or to join in secular revelry to be just like them, but instead to develop the fruit of the Spirit so that we are friendly, reliable, hardworking, and genuinely caring for others (see 2 Cor. 5:18–21). Good character and good deeds are the best way to break stigma. I suspect that if my student brought doughnuts, bagels, or cookies to share with everyone in the lab a few times a week over the summer, along with being friendly and doing excellent work, then the group would look for student from Biola to work with them again next summer.[13] After all, doing good is what Peter enjoins us to do:

> Keep your conduct among the Gentiles honorable, so that when they speak against you as evildoers, they may see your good deeds and glorify God on the day of visitation. . . . For this is the will of God, that by doing good you should put to silence the ignorance of foolish people. Live as people who are free, not using your freedom as a cover-up for evil, but living as servants of God. (1 Pet. 2:12, 15–16)

6. CHRISTIANS NEED TO BE A POSITIVE INFLUENCE IN THE ETHICS OF SCIENCE

In addition to personal character, Christians in the sciences should work to influence the directions of research, medicine, and technology for the greatest good. Like most human endeavors, science itself is generally ethically neutral, but any particular application can be good or evil. A knife can either heal or kill. Nuclear energy can make electricity or bombs.

But there are situations where the research topics we choose or the means by which we go about gaining knowledge raise ethical problems. Today most scientists recognize that it was unethical for Nazi doctors to torture prisoners in Hitler's concentration camps,

[13] The apologetics group Stand to Reason (www.str.org) specializes in training people to give wise and humble answers when challenged, along with pointers on character development. I highly recommend their work. By the way, my student reported the following fall that she had a great time working in the lab and was well liked.

thus the physiological data they gained from freezing or starving their victims is not used. Modern stem cell research offers a less dramatic but similar dilemma: specimens can be taken either from embryos and aborted fetuses or from neonatal and adult tissues. Given the moral issues, it would seem appropriate for Christians to encourage the use of cells which are not obtained at the cost of a human life. Human cloning is another potential technology that raises questions—should we clone someone in order to harvest healthy organs from the clone as some have suggested? However we think the soul is imparted to us, making a clone to serve as a sacrificial donor seems odd from a Christian perspective. I've often heard the attitude, "If we can do it, we should," regarding science and technology; yet I don't think that justifies selling nuclear weapons at the grocery store. We need to consider the consequences of our actions.

The Christian tradition here is simple and clear: we should strive to do good works. Jesus says this best,

> You are the light of the world. A city set on a hill cannot be hidden. Nor do people light a lamp and put it under a basket, but on a stand, and it gives light to all in the house. In the same way, let your light shine before others, so that they may see your good works and give glory to your Father who is in heaven. (Matt. 5:14–16)

Notice that "let your light shine" parallels "see your good works." As Christians in the sciences, we should focus our efforts on research and concerns that will help others. While we don't think that technology will save us, as much of the world does, it certainly can relieve pain and suffering, and can promote human flourishing. While medicine obviously comes to mind (Christians founded the first public hospitals, by the way), improving our food, water, and energy supplies (with an eye to the wise stewardship of our resources), or designing safer buildings and vehicles, are all channels for good works. And given our fallen human nature, so is the development of weapons for defensive and deterrent purposes.

Unfortunately, much of our technology and research today is driven by the profit motive instead of good works. While these goals can overlap, there are many situations where they do not: critics have noted that the pharmaceutical industry often focuses on developing expensive drugs which patients need to take for the rest of their lives, rather than on simple and cheap remedies. Or once a blockbuster drug goes off-patent, the company often hypes a new, "improved" version, which doctors sometimes later discover is less effective than the old one. Or research on curing major third-world diseases is not done because any cure which is developed would not be profitable given the market for it. Of course, often there are complicating social factors: drug companies were loath to develop new and better vaccines because of the threat of lawsuits from extremely rare adverse reactions. It doesn't make sense to produce a five-dollar vaccine if there's a one-in-a-million chance that you will have to deal with billion-dollar legal settlements.[14]

I am not intending to single out pharmaceutical companies here, because if you look under the rug of automakers or almost any industry and scientific/technological endeavor, you will find many examples where the profit motive has led them astray. But if the profit motive, which is a practical economic necessity due to man's fallen nature, were tempered with the Christian emphasis on doing good and being honest in all things, it would greatly help human flourishing.

CONCLUSION

It is impossible to study the world without a set of assumptions. These assumptions are often part of the culture in which we live and are, therefore, as difficult to isolate and examine as it is for a

[14] This particular situation has been partially remedied by the 1986 National Childhood Vaccine Injury Act, but because damage claims are difficult to prove, few are compensated, giving critics the impression that the law shields manufacturers and doesn't help the injured enough.

fish to isolate itself from water. But if we put our cultural myths into proper historical perspective, we see that the endeavor which we call "science" today developed within the Christian tradition in Europe because a biblical view of nature gave us a good working set of assumptions.

But because of the influence of Aristotelian views, these assumptions were not entirely correct. Galileo argued with the academics in his day for the need to study nature itself and not base our understanding of the world purely on philosophical principles or on a particular Aristotelian-friendly interpretation of a biblical text. Only by carefully examining the data from nature and our own assumptions about it can we make progress in understanding both God's Word and his world.

I sometimes feel like a modern Galileo: but instead of Aristotelian dogma, the sciences today are trapped in a naturalistic dogma that *by definition* rules out the most obvious interpretation of the physical data, especially in the areas of astronomy and biology that address origins. Sadly, the naturalistic assumptions in our culture are powerful enough even to prevent many Christians from seeing the hand of God in the world around us. Yet I am encouraged that, as we have noted, in candid moments a number of prominent scientists recognize that a secular, naturalistic worldview offers no rational basis for their work. Only when we recognize that the universe is the work of a divine mind and that we have some ability to understand it because we are made in his image, does science—and even life itself—make sense, turning a secular farce into a divine tribute. C. S. Lewis summarizes this point well:

> The proof or verification of my Christian answer to the cosmic sum is this. When I accept Theology I may find difficulties, at this point or that, in harmonizing it with some particular truths which are imbedded in the mythical cosmology derived from science. But I can get in, or allow for, science as a whole. Granted that Reason is prior to matter and that the light of the

> primal Reason illuminates finite minds, I can understand how men should come, by observation and inference, to know a lot about the universe they live in. If, on the other hand, I swallow the scientific cosmology as a whole, then not only can I not fit in Christianity, but I cannot even fit in science. If minds are wholly dependent on brains, and brains on biochemistry, and biochemistry (in the long run) on the meaningless flux of the atoms, I cannot understand how the thought of those minds should have any more significance than the sound of the wind in the trees. And this is to me the final test. This is how I distinguish dreaming and waking. When I am awake I can, in some degree, account for and study my dream. The dragon that pursued me last night can be fitted into my waking world. I know that there are such things as dreams; I know that I had eaten an indigestible dinner; I know that a man of my reading might be expected to dream of dragons. But while in the nightmare I could not have fitted in my waking experience. The waking world is judged more real because it can thus contain the dreaming world; the dreaming world is judged less real because it cannot contain the waking one. For the same reason I am certain that in passing from the scientific points of view to the theological, I have passed from dream to waking. Christian theology can fit in science, art, morality, and the sub-Christian religions. The scientific point of view cannot fit in any of these things, not even science itself. I believe in Christianity as I believe that the Sun has risen, not only because I see it, but because by it I see everything else.[15]

The Christian intellectual tradition in the sciences strives to seek truth, seeks harmony in God's dual revelations of himself in Scripture and creation, and expects our ongoing study of God's creation to reveal his handiwork more and more. Fundamental to the tradition, however, is a proper understanding of God as the Creator of the universe. Can we grasp an appreciation for the Being who made billions of galaxies, and who crafted the laws and constants with an elegance and precision so that the "just right" physics and chemistry could exist? Can we catch a glimpse of the

[15] C. S. Lewis, "Is Theology Poetry?" in *The Weight of Glory* (New York: HarperOne, 2009), 138–40.

Genius who designed and programmed chemistry to make living creatures, especially complex ones like us, and who placed them on the fragile, delicate, yet protected speck of dust we call Earth? If we can but grasp this vision, we have a framework for our own significance and for the pursuit of science. As God's redeemed children, we can explore and marvel at our Father's handiwork and harness it for his glory and the betterment of others.

QUESTIONS FOR REFLECTION

1. In the first section of chapter 1, we surveyed key elements of the Christian foundation for doing science. Which one struck you as the most interesting and why?

2. Before reading this book, how aware were you of the Christian heritage that modern science has? What might you do in the future to strengthen your background in the history of science?

3. How does our culture's default position of naturalism influence your life as a Christian? What are some steps you can take to overcome it?

4. What are the goals of science? Which ones impact the Christian worldview? Which ones flow from a Christian worldview?

5. As noted in chapter 5, philosophers suggest that science should either deliberately limit its explanations or else allow for non-naturalistic explanations. Which do you think is the more feasible approach and why?

6. Why do you think that the discipline of biology is one of the more challenging fields for Christians today?

7. Many scholars have noted that the perceived tensions between Christianity and science are actually due to the religious tensions between Christianity and naturalism. Discuss the significance of this insight.

TIMELINE

350s BC	Plato and Aristotle. Classical Western philosophy formulated.
AD 150s	Ptolemy publishes the *Almagest*, a formal presentation of the geocentric solar system.
1100s	Universities start throughout Western Europe to train professional lawyers, government workers, physicians, and theologians.
1543	Nicolaus Copernicus publishes his heliocentric model, *On the Revolutions of the Celestial Spheres*.
1570s	Tycho Brahe makes the most accurate naked-eye, pre-telescopic astronomical observations.
1609	Johannes Kepler publishes *The New Astronomy*, using elliptical planetary orbits.
1610	Galileo Galilei publishes *The Sidereal Messenger*, the first astronomical telescope observations.
1620	Francis Bacon articulates the scientific method.
1644	René Descartes publishes *Principles of Philosophy*.
1662	Robert Boyle—ideal gas laws.
1687	Isaac Newton publishes *Principia*, laws of motion and gravity.
1750s	Voltaire active; the Enlightenment movement publishes *Encyclopaedia* "to change the way people think."
1831	Michael Faraday—electromagnetic induction.
1859	Charles Darwin publishes *On the Origin of Species*.
1873	James Clerk Maxwell—unified theory of electromagnetism.
1929	Edwin Hubble observes galaxy recession speeds, proposes expanding universe.
1953	Francis Crick and James Watson—structure of DNA.
1964	Anro Penzias and Robert Wilson detect cosmic microwave background radiation, evidence for the big bang theory.
1973	Brandon Carter proposes anthropic principle.

GLOSSARY[1]

Active principle. A force or power that can act on matter, which can include God, divine beings and agents, mathematical and moral truths, human intellect, or gravity.

Aristotle. (384–322 BC) Greek philosopher who wrote on logic, biology, rhetoric, and politics, and whose surviving works strongly influenced centuries of Western thought from about AD 1100–1600s.

Counter-Reformation. (1560–1648) Reforms made by the Roman Catholic Church in response to charges made by Luther and other Protestants, which included financial, political, educational, and doctrinal changes and clarifications.

Darwinism. *See* Macroevolution.

Deism. The belief that God created the world but does not interfere with its subsequent operation in any way, nor has he revealed himself through any supernatural means.

Demiurge. A subordinate god who crafted the world, distinct from the supreme god. In gnostic thought, he was seen as imperfect and evil.

Determinism. The notion of absolute physical cause and effect, where any situation is completely determined by prior conditions.

DNA. Deoxyribonucleic acid is a long, double-stranded molecule which contains the biological instructions necessary for making all living organisms.

Enlightenment. Intellectual and cultural movement from about 1650–1800, which emphasized reason, mathematics, and experimentation while despising tradition.

Epicycles. A system of "circles attached to larger circles" that Ptolemy used to calculate the position of planets in the sky.

Geocentric. A model of the universe where the earth is at the center, and the sun, moon, and planets orbit around it. Often called the "Ptolemaic" system.

Heliocentric. Model of the universe where the sun is at the center and the planets orbit it.

Immanence. God is present in, but not part of, the physical world, and he relates to it intimately.

Intelligent design. The proposal that the physical and biological worlds show patterns that are best explained as the products of an intelligent agent rather than undirected natural processes.

Macroevolution. Evolution on a large scale, focusing on the major changes between organisms that involve the development of new organs, body plans, and structures. Also called "Darwinism." Compare with "microevolution." Creationists accept microevolution but question whether small-scale change can be extrapolated as evidence for and a mechanism for macroevolution, apart from God's guidance and/or intervention.

Messenger RNA (mRNA). A chain of ribose nucleotide molecules that carry genetic information from the cell's DNA "cookbook" to the ribosomes, which read the RNA sequence and construct a specific protein from the encoded message.

[1] *The Stanford Encyclopedia of Philosophy* (plato.stanford.edu) is an excellent resource for digging deeper on the philosophical terms in this glossary.

Metaphysics. Investigation about the nature of reality; of what actually exists and why.

Methodological naturalism. An approach which attempts to explain all physical phenomena on the basis of natural causes, as opposed to supernatural or spiritual ones. This position does not presume that *metaphysical naturalism* is true (*see* Naturalism), but adopts it only for explaining the behavior of physical objects.

Microevolution. Small-scale changes in the features of a single species over time, often in response to environmental factors.

Naturalism. More precisely called *metaphysical naturalism*, the idea that the natural world is all that exists, and there is no supernatural.

Neo-Darwinism. The modern version of Darwinism, which proposes that the gradual changes in organisms over time occur because of genetic mutations. Darwin proposed natural selection, but did not know how variation among individual organisms occurs or is transmitted to offspring. Macroevolution, Neo-Darwinism, and Darwinism are commonly used as synonyms.

Nucleotide. A class of organic molecules that form the "building blocks" for constructing long chains of DNA or RNA.

Positivism. A philosophy developed by Auguste Comte in the nineteenth century, who argued that the only valid knowledge is that which can be found through logic, mathematics, and careful experiment. Thus science is the only path to truth, and any other forms of or routes to knowledge are rejected.

Providence. God's guidance and care over the events of the world, without requiring distinct intervention.

Reductionism. A philosophy which argues that complex systems can be fully explained by understanding how the laws of physics and chemistry act on the sum of their parts.

RNA. Ribonucleic acid is a family of information-carrying molecules which perform a wide range of control functions in cells.

Scientism. The view that science offers the only means to gain truth, and that all other disciplines, especially religion, are worthless.

Teleology. The belief that purpose and design in the world are real and can be studied.

Theism. Belief in a personal God who created and actively rules the world, and who can reveal himself to people.

Transcendence. God is completely beyond and independent of the physical universe and its laws.

RESOURCES FOR FURTHER STUDY

The following books are organized with the reader's background and specific areas of interest in mind, and then annotated to help give the reader a sense of direction for digging deeper.

INTRODUCTORY

1. Nancy R. Pearcey and Charles B. Thaxton. *The Soul of Science: Christian Faith and Natural Philosophy*. Wheaton, IL: Crossway, 1994.

   A more detailed presentation of many of the themes I have outlined in this book; a good starting point for those who want to dig deeper.

2. John C. Lennox. *God's Undertaker: Has Science Buried God?* 2nd ed. Oxford: Lion, 2009.

   The best brief introduction to the contemporary debate between Christianity and science that I am aware of.

3. Benjamin Wiker and Jonathan Witt. *Meaningful World: How the Arts and Sciences Reveal the Genius of Nature*. Westmont, IL: IVP Academic, 2006.

   An excellent introduction to the common features which all creative works share. Starting with the genius of Shakespeare, the authors' unique multidisciplinary approach gives insights into the Designer of the world around us.

4. Dan Graves. *Scientists of Faith: 48 Biographies of Historic Scientists and Their Christian Faith*. Grand Rapids, MI: Kregel, 1996.

   A convenient and inspiring resource for role models and examples of Christians in science.

5. Charles E. Hummel. *The Galileo Connection: Resolving Conflicts between Science and the Bible*. Westmont, IL: InterVarsity, 1996.

   Extremely readable account of Galileo and other early scientists' lives, beliefs, and accomplishments. Helps dispel myths about Galileo's trial and traces the influence of Christian thinking in the sciences.

INTERMEDIATE

1. J. P. Moreland. *Christianity and the Nature of Science: A Philosophical Investigation*. 2nd ed. Grand Rapids, MI: Baker, 1999.

   Considered one of the classic treatments on the philosophy of science from a Christian perspective. A very readable place to start in this discipline.

2. Gary B. Ferngren. *Science and Religion: A Historical Introduction*. Baltimore, MD: Johns Hopkins University Press, 2002.

   An anthology of accessible but professional articles on a range of key topics.

3. J. P. Moreland. *Kingdom Triangle: Recover the Christian Mind, Renovate the Soul, Restore the Spirit's Power*. Grand Rapids, MI: Zondervan, 2007.

   An excellent work on integrating the intellectual and spiritual Christian life.

SPECIALIZED

1. Christopher B. Kaiser. *Creational Theology and the History of Physical Science: The Creationist Tradition from Basil to Bohr*. Leiden: Brill, 1997.

   The "gold standard" scholarly assessment of the influence of Christianity on the rise of classical and modern science.

2. Stephen C. Meyer. *Signature in the Cell: DNA and the Evidence for Design*. New York: HarperOne, 2010.

   Surveys contemporary naturalistic origin-of-life models in detail and notes their weaknesses. Very readable, engaging, and respectful of opposing positions.

3. Stephen C. Meyer. *Darwin's Doubt: The Explosive Origin of Animal Life and the Case for Intelligent Design*. New York: HarperOne, 2013.

   Explores the Cambrian explosion and its significance for macroevolutionary theories. A readable survey of developmental biology that respectfully shows the inability of naturalistic models to explain the big questions concerning origins in biology today.

4. Fazale Rana. *The Cell's Design: How Chemistry Reveals the Creator's Artistry*. Grand Rapids, MI: Baker, 2008.

   A survey of cell biology, noting the design features and the immense level of complexity that defy a naturalistic origin.

5. Fazale Rana and Hugh Ross. *Who Was Adam? A Creation Model Approach to the Origin of Man*. Colorado Springs, CO: NavPress, 2005.

   An overview of the scientific data which relate to the human origins question, combined with an excellent integration of that data to the traditional account of a created, historical Adam and Eve.

6. Fazale Rana. *Origins of Life: Biblical and Evolutionary Models Face Off*. Colorado Springs, CO: Navpress, 2004.

   Earlier and more technical than *Signature in the Cell*, a good survey of the problems which naturalistic origin-of-life theories face.

GENERAL INDEX

SCRIPTURE INDEX